UNITEXT for Physics

More information about this series at http://www.springer.com/series/13351

Piero Olla

An Introduction
to Thermodynamics
and Statistical Physics

 Springer

Piero Olla
ISAC-CNR
Cagliari
Italy

ISSN 2198-7882   ISSN 2198-7890   (electronic)
ISBN 978-3-319-36049-2   ISBN 978-3-319-06188-7   (eBook)
DOI 10.1007/978-3-319-06188-7

Springer Cham Heidelberg New York Dordrecht London

Printed on acid-free paper

Springer is part of Springer Science+Business Media (www.springer.com)

Contents

Chapter 1
Introduction

One of the fundamental problems of physics is the determination of the properties of macroscopic systems from those of their microscopic constituents. In principle, such properties could be obtained from an exact knowledge of microscopic state of the system. Such an amount of information, however, beyond being impossible to obtain, is also absolutely redundant (we do not need to know position and velocity of all the molecule in a gas, to determine e.g. its pressure). Systems that are more or less complicated, may require a large number of variable to be described in a decent way, but the general principle that a full microscopic description is not necessary, remains valid.

The passage to a macroscopic level of description, clearly implies a loss of information, and it is not obvious that such operation is always possible. A first condition is that the interactions among the microscopic constituents is simple (as it is in the case of the molecules in a gas; not so, say, in the case of the neurons in a human brain). Such simplicity guarantees that the macroscopic variables depend in relatively simple way on their microscopic counterparts (typically through sums). Another condition that seems to play a crucial role is the existence of a state of thermodynamic equilibrium, towards which the system tends, irrespective of initial conditions, if left on its own. Loss of memory of initial conditions is in fact one of the facets of the irreversible character of the relaxation to equilibrium. We could consider subdivisions of the system in smaller and smaller parts, down to the microscopic scale. This would lead us to expect a chaotic character (in some sense) of the microscopic dynamics. The point is controversial. What is clear is that loss of memory of initial conditions would imply a loss of information in the passage from micro to macro, that is not due merely to coarse-graining, and has dynamical origin. Such ideas extend far from equilibrium, with the concept of thermal equilibrium being replaced by that of non-equilibrium steady state, in which macroscopic fluctuations, nonlinear effects and chaos, become the dominant factors.

The fact that macroscopic variables can be obtained as sums (or in general simple functions) of the microscopic variables, suggests that they could be interpreted in statistical sense. An example is the internal energy of a monoatomic gas $E = (3/2)NKT$,

© Springer International Publishing Switzerland 2015
P. Olla, *An Introduction to Thermodynamics and Statistical Physics*,
UNITEXT for Physics, DOI 10.1007/978-3-319-06188-7_1

in which $(3/2)KT$ is the mean kinetic energy of the molecules. The concepts of probability and statistics play therefore a central role. Basically, two approaches exist: kinetic theory, associated with the names, among others, of J.C. Maxwell and L. Boltzmann, and equilibrium statistical mechanics, associated with the name of W. Gibbs. Although application of these theories extend nowadays well beyond the realm of physics, their original use was in the study of thermodynamic systems. Simplifying a lot, one can state that kinetic theory is more concerned with non-equilibrium situations, while statistical mechanics could be seen as a body of techniques for the derivation of the thermodynamics of physical systems. In both cases, central aspect is the probabilistic treatment of the microscopic degrees of freedom of the system.

An essential role in establishing the connection between microscopic and macroscopic worlds, is played by entropy. Entropy is sometimes described as a measure of disorder. Unfortunately, this is a rather ambiguous definition, as it is not clear from the start what is intended in general with the term disorder. Also, characterizing the processes taking place, say, in a biologic cycle, as an increase of disorder, seems somewhat limited, even though such processes are always associated with an increase of entropy. In fact, the concept of entropy takes different meanings depending on the context. All of them can be brought back, however, to the amount of information lost in the passage to a macroscopic description of the system. In the notes that follow, particular attention will be given to the different roles played by entropy in thermodynamics, kinetic theory and statistical mechanics, highlighting in all cases its meaning from the point of view of information theory (Shannon entropy).

These notes contain a presentation at an introductory level of kinetic theory and statistical mechanics. The presentation is partly inspired by that in classical textbooks, such as "Statistical Physics", by L. Landau and "Thermodynamics and an introduction to thermostatistics", by H.B. Callen. Rather than providing a comprehensive presentation of techniques and applications, attention has been given to the basic tools, illustrating their application with simple systems and toy models. Some flavor of the wide range of application of these techniques is nevertheless given. Problems ranging from population dynamics, to atmospheric physics, to chemistry, are taken into consideration, trying to keep an eye, in all situations, on the micro-macro issue, and on how macroscopic properties arise from the microscopic dynamics. A preference, in the presentation, has been given to classical systems, which allows a deeper insight into the construction of tricky concepts such as those of microstate and indistinguishable particles.

The book assumes some familiarity with basic concepts of thermodynamics and probability, but not beyond what is commonly obtained in the undergraduate curriculum. The idea is that of an intermediate textbook, providing the necessary tools of probability theory and thermodynamics, for later study of statistical mechanics in specialized courses at the graduate level. At the same time, a bird's eye view is given on arguments (kinetic theory, the theory of fluctuation) that often are disregarded in the main curriculum courses.

Great attention is given to the role of entropy, as a measure of the loss of information in passing from a microscopic to a macroscopic description. As regards the part on kinetic theory, importance is given to toy models to explain concepts such as the problem of closure, and techniques such as that of moment equations. The part on fluctuations contains a brief account of recent results on the relations between entropy production and time irreversibility in mesoscopic systems.

Chapter 2
Review of Probability and Statistics

2.1 Basic Definitions

Probability is one of those familiar concepts that turn out to be difficult to define formally. The commonly accepted definition, is the axiomatic one due to Kolmogorov, that provides the minimal set of properties that a probability must satisfy, but does not say anything about what it represents. In fact, we know perfectly well what probability should represent: a measure of our expectation of the outcomes of a given experiment. Unfortunately, this is not enough to build a theory (for instance, how do we define the concept of expectation?).

The axiomatic definition of probability is obtained from elementary concepts in the following way.

- Let us call the set Ω of the possible outcomes of a given experiment the **sample space** for that experiment. Examples: the set $\Omega = \{1, 2, 3, 4, 5, 6\}$ of the results of a dice roll; the set $\Omega = \mathbf{R}^3$ of the possible velocities of a molecule in a gas. A quantity, such as the velocity of the molecule, whose sample space is continuous, is called a **random variable**.
- A set A of possible results of an experiment is called an **event**. An event is by definition a subset of Ω; it must be mentioned, however, that in the case of a continuous sample space, not all the subsets $A \subset \Omega$ are admissible; we shall not discuss further this topic. See e.g. [W. Feller, An introduction to probability theory and its applications (Wiley and Sons, 1968), Vol. II, Sect. I.11]. Two events A and B are said to be **mutually exclusive** if their intersection is empty $A \cap B = \varnothing$.
- The **probability** of an event a is a number $P(A)$, $0 \leq P(A) \leq 1$, with the properties $P(A \cup B) = P(A) + P(B)$, in the case A and B are mutually exclusive, and $P(\Omega) = 1$.

Not bad with regard to economy, for a definition. It can be verified that the two axioms lead to a probability that has all the properties suggested by our intuition (e.g. that $P(\varnothing) = 0$). Notice that the expression $P(A \cup B)$ is nothing but a stenographic notation

© Springer International Publishing Switzerland 2015
P. Olla, *An Introduction to Thermodynamics and Statistical Physics*,
UNITEXT for Physics, DOI 10.1007/978-3-319-06188-7_2

to indicate the probability that the experiment produces an outcome belonging to either A or B.

In the case of a random variable x, events can be obtained, in general, as union and intersection of intervals of $\mathbf{R}$. We can then define a **probability density function** (PDF) $\rho(x)$ through the formula

$$P([a, b]) = \int_a^b \rho(x)\mathrm{d}x. \tag{2.1}$$

Once a certain system of units is selected, a physical event will correspond, typically, to an interval of values of the physical quantity expressed in those units. In other units, the corresponding interval would of course be different. The PDF of the same physical quantity in different units will be different as well. For instance, a physical event $A \equiv [a, b]$ in the variable x, will transform into $[2a, 2b]$, in the variable $y = 2x$. From Eq. (2.1), the probability $P(A)$ will be expressed therefore in terms of the PDF for the variables x and y:

$$P(A) = \int_a^b \rho_x(z)\mathrm{d}z = \int_{2a}^{2b} \rho_y(z)\mathrm{d}z \tag{2.2}$$

(subscript x and y label the random variables; z are their values). Considering infinitesimal intervals, would give us in Eq. (2.2): $\rho_x(z)\mathrm{d}z = 2\rho_y(2z)\mathrm{d}z$, from which we obtain the formula for the change of variables in PDF's:

$$\rho_x(x) = \rho_y(y(x)) \left| \frac{\mathrm{d}y(x)}{\mathrm{d}x} \right|. \tag{2.3}$$

We can associate probabilities to isolated points in a continuous sample space. A non-zero probability for such events, will produce a Dirac-delta component in the corresponding PDF. For instance:

$$\rho(x) = a\delta(x) + b\theta(x)\exp(-x), x \in \mathbf{R}, \tag{2.4}$$

and we have of course $P(0) = a$. In order to have $P(\Omega) = 1$, we must impose $b = 1 - a$.

2.1.1 Joint, Conditional and Marginal Probabilities

Given two events A and B, we call the **joint probability** of A and B, the probability of the intersection $P(A \cap B)$. Important example: in the case A and B are, respectively, a vertical and a horizontal slice of a domain $\Omega \subset \mathbf{R}^2$, $A = [x, x + \mathrm{d}x] \otimes \Omega_y(x)$, and $B = \Omega_x(y) \otimes [y, y + \mathrm{d}y]$, with $\Omega_y(x)$ the set of the y contained in Ω at horizontal position x, and $\Omega_x(y)$ the set of the x contained in Ω at vertical position y. We define the **joint PDF** $\rho_{xy}(x, y)$ of the values x and y of the random variables x and y, from the relation

$$P(A \cap B) = P([x, x + dx] \otimes [y, y + dy]) := \rho_{xy}(x, y)dxdy. \qquad (2.5)$$

(In the following we shall put subscripts on the PDF's only when strictly necessary). Of course, we can define objects that are half-way between probability and PDF (e.g. integrating $\rho(x, y)$ with respect to x in an interval A; thus $\rho(A, y)$ is a probability with respect to A, and a PDF with respect to y).

Note The formula for the change of variable Eq. (2.3) can be generalized to higher dimension. If $\mathbf{y} = \mathbf{y}(\mathbf{x})$: $\rho_{\mathbf{x}}(\mathbf{x}) = \rho_{\mathbf{y}}(\mathbf{y}(\mathbf{x}))|J_{\mathbf{y}}(\mathbf{x})|^{-1}$, where $J_{\mathbf{y}}(\mathbf{x}) = \det(\partial_{\mathbf{x}}\mathbf{y}(\mathbf{x}))$ is the Jacobian determinant of the transformation. Notice that we use here the vector notation $\mathbf{x} = (x_1, x_2, \ldots)$ and $\mathbf{y} = (y_1, y_2, \ldots)$ to indicate two sets of vector components that represent the same vector. For instance, $\mathbf{x}$ and $\mathbf{y}$ may represent polar and cartesian coordinates in the plane, $\mathbf{x} = (r, \phi)$, $\mathbf{y}(\mathbf{x}) = (r\cos(\phi), r\sin(\phi))$, so that $\rho_{r,\phi}(r, \phi) = r\rho_{y_1,y_2}(r\cos(\phi), r\sin(\phi))$.∎

We introduce next the concept of **conditional probability** of an event A conditional to an event $B \neq \varnothing$:

$$P(A|B) = P(A \cap B)/P(B). \qquad (2.6)$$

The conditional probability $P(A|B)$ can thus be seen as the probability of A obtained considering as sample space B. We can generalize again to PDF's:

$$\rho(x|y) = \rho(x, y)/\rho(y); \quad P([x, x + dx]|y) = \rho(x|y)dx. \qquad (2.7)$$

(prove this last result from the definition Eq. (2.5)).

Two events are said to be **statistically independent**, if $P(A \cap B) = P(A)P(B)$. The concept of conditional probability allows us to get an intuitive grasp of this definition. If A and B are statistically independent (and are not empty), we have in fact

$$P(A|B) = P(A); \qquad P(B|A) = P(B). \qquad (2.8)$$

In other words, conditioning by a statistically independent event, does not modify the probability. Similarly, in the case of a PDF: the condition $\rho(x|y) = \rho(x)$ tells us that the occurrence of x, as outcome of an experiment, is independent of the fact that, simultaneously, y takes a particular value. If $\rho(x, y) = \rho(x)\rho(y) \, \forall x, y$, we shall say that the random variables x and y are statistically independent.

At last, we notice the following relation, that descends directly from the definition Eq. (2.5):

$$\rho(x) = \int_{\Omega_y(x)} \rho(x, y)dy; \quad \rho(y) = \int_{\Omega_x(y)} \rho(x, y)dx. \qquad (2.9)$$

In this context, the two PDF's $\rho(x)$ and $\rho(y)$ are called **marginal**.

2.1.2 The Concept of Average

We define the average of a function f of a random variable x, through the relation:

$$\langle f \rangle = \int \rho(x) f(x) dx, \tag{2.10}$$

which, in the case of a discrete random variable, becomes, using Eq. (2.4):

$$\langle f \rangle = \sum_i P(x_i) f(x_i).$$

We can take averages, starting from a conditional PDF, and the result is called a **conditional average**:

$$\langle f | y \rangle = \int \rho(x) f(x | y) dx, \tag{2.11}$$

which, being y still random, will be itself a random quantity.

A little review of notation and terminology. Typically, $\mu_x \equiv \langle x \rangle$ is called the mean of the PDF, and $\sigma^2 = \langle (x - \mu_x)^2 \rangle = \langle x^2 \rangle - \langle x \rangle^2$, the variance of the PDF (σ_x is called standard deviation or RMS of x). The average $\langle x^n \rangle$ is called nth **moment** of the PDF. The quantity $\langle f g \rangle$ is called **correlation** of f and g. Important case:

$$C_{xy} = \langle xy \rangle = \int \rho(x, y) xy dx dy. \tag{2.12}$$

Notice that statistical independence of x and y implies $C_{xy} = 0$, but the vice versa is not true (find a counterexample).

An interesting object, of which to take averages, is the **indicator function** of an event A:

$$\delta_A(x) = \begin{cases} 1 & \text{if } x \in A, \\ 0 & \text{otherwise.} \end{cases} \tag{2.13}$$

We find:

$$\langle \delta_A \rangle = \int_A \rho(x) dx = P(A). \tag{2.14}$$

The indicator function of a random variable x, is just the Dirac delta:

$$\delta_{\bar{x}}(x) \equiv \delta(x - \bar{x}) \Rightarrow \langle \delta_{\bar{x}} \rangle = \rho(\bar{x}). \tag{2.15}$$

Equations (2.13–2.14), and in some sense also Eq. (2.15), are the starting point for the statistical determination of probabilities (and PDF's) from the concept of frequency.

2.1.3 Characteristic Function

Another quantity of which to take averages, is the exponential of a random variable: the so called **characteristic function**

$$Z(j) = \langle \exp(ijx) \rangle = \int dx \, \rho(x) \exp(ijx), \qquad (2.16)$$

that is the Fourier transform of the PDF $\rho(x)$.

We notice a number of important properties.

The PDF $\rho(x)$ and the characteristic function $Z(j)$ are, respectively, the average of the indicator function δ_x, and of the exponential $\exp(ijx)$, where the second can be written trivially as the Fourier transform of the Dirac delta:

$$\exp(ijx) = \int dx' \delta_x(x') \exp(ijx').$$

We can invert the Fourier transform in Eq. (2.16), to obtain again the PDF:

$$\rho(x) = \int \frac{dj}{2\pi} Z(j) \exp(-ijx),$$

which corresponds to taking the average of the Fourier integral representation of the Dirac delta:

$$\delta_x(x') \equiv \delta(x - x') = \int \frac{dj}{2\pi} \exp(ij(x' - x)).$$

The most important property of the characteristic function, however, is that its derivatives at $j \to 0$, are just the moments of ρ (provided they exist):

$$\lim_{j \to 0} \frac{d^n Z(j)}{dj^n} = \lim_{j \to 0} \int dx \, (ix)^n \rho(x) \exp(ijx) = i^n \langle x^n \rangle \qquad (2.17)$$

(notice that, from normalization of ρ, we shall always have $Z(0) = 1$). Existence of all the moments of a PDF, will imply analyticity of $Z(j)$ in $j = 0$, with the moments of ρ providing the coefficients of the Taylor expansion of Z.

2.2 Entropy and Information

We have seen that the probability of an event, parameterizes our expectation that particular event occurs or not. A question that is worth to ask is how much information on the outcome of an experiment, is gained by knowing the distribution of its possible results. It is rather clear that, given a coin toss experiment, we know more about its outcome if the probability distribution is $P(head) = 1$, $P(tail) = 0$, rather than if $P(head) = P(tail) = 1/2$. Shannon found a natural way to parameterize this information content, introducing an entropy, that is neither that of thermodynamics, nor that of statistical mechanics, but that has a lot to do with both. As we shall see, the entropy introduced by Shannon, does not parameterize the content of information of a distribution, instead, that of "ignorance", i.e. the amount of additional information, that should be provided to know with certainty the result of an experiment.

We point out that the concept of information depends a lot on the kind of experiment. In the case of the coin toss experiment, the question may seem obvious; not so in the case of a random variable. The fact is that we cannot measure a random variable with infinite precision; what we are able to "measure" are only events. We need in some way to discretize Ω. We define a **partition** of Ω, as a collection of mutually exclusive events A_i, such that their union is Ω itself (see Fig. 2.1):

$$\mathscr{P} = \{A_i, i = 1, \ldots; \ A_i \cap A_j = \varnothing \text{ if } i \neq j; \ \cup_i A_i = \Omega\}. \tag{2.18}$$

We define the **Shannon entropy** of the distribution P, referred to the partition $\mathscr{P}$ of Ω, as:

$$S[P] = -\sum_i P(A_i) \ln P(A_i) = -\langle \ln P \rangle, \tag{2.19}$$

where the notation $S[P]$ is a shorthand for $S(\{P_i, i = 1, \ldots\})$. We see at once that the definition works smoothly in the case of the coin toss experiment, in which, trivially, $\mathscr{P} \equiv \{head, tail\}$. If $P(head) = 1$ (and $P(tail) = 0$) or $P(tail) = 1$ (and $P(head) = 0$), as in the case of a loaded coin, we will find in fact $S[P] = 0$ (we know the result: minimum ignorance, implying minimum entropy, as $0 \leq P \leq 1 \Rightarrow \ln P \geq 0 \Rightarrow S[P] \geq 0$). On the other extreme, $P(head) = P(tail) = 1/2$, we have instead $S[P] = \ln 2$ (maximum ignorance).

Actually, things work *very* well.

The key point is that entropy is in the form

$$S[P] = \sum_i g(P_i),$$

with $g(P)$ a **convex** function of the argument. In other words, the following relation is satisfied:

$$\overline{g(P_i)} \leq g(\overline{P}), \tag{2.20}$$

where $\overline{x} = \frac{1}{N} \sum_{i=1}^{N} x_i$ is the arithmetic mean of the x_i's. The situation is illustrated in Fig. 2.2, in the case of two events A_1 and A_2 with probabilities respectively P_1 and P_2. Their contribution to the entropy is $g(P_1) + g(P_2) = 2\overline{g(P)}$. If these probabilities were equal, all the others, $P_i, i = 3, \ldots$, maintaining their original values, we would have $P(A_1) = P(A_2) = (P_1 + P_2)/2 = \overline{P}$. In that case, the contribution to entropy would be $2g(\overline{P}) \geq 2\overline{g(P)}$. From here, we can conclude that the entropy $S[P]$ will become maximum, when all the events in the partition $\mathscr{P}$ are equiprobable.

Fig. 2.1 Partition of a
two-dimensional domain Ω
(sample space)

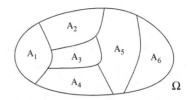

Fig. 2.2 Graphic representation of Eq. (2.20) in the case $N = 2$

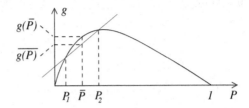

The information that must be supplied will be larger, again, if the number of outcomes of the experiment increases. This is what happens if we refine the partition, e.g. if we split event A_6 in Fig. 2.1, into two disjoint events A_{61} and A_{62}. We see that, also in this case, entropy behaves as expected, increasing in response to refinement. In fact, the contribution to entropy, from A_{61} and A_{62}, will remain the same as before, only if $P(A_{61}) = P(A_6)$ and $P(A_{62}) = 0$, or vice versa. Otherwise, due to convexity of g: $g(A_6) < g(A_{61}) + g(A_{62})$.

Another property that we should expect from entropy, if it is really a measure of an information content, is that its value, referred to two independent systems, should be the sum of the entropies referred to each system individually. Thus, if systems A and B are described by physical variables, with domains Ω_A and Ω_B, and all the elements of the partitions $\mathscr{P}_A = \{A_1, A_2, \ldots\}$, $\mathscr{P}_B = \{B_1, B_2, \ldots\}$, are statistically independent, $P(A_i, B_j) = P_A(A_i)P_B(B_j)$, we should have: $S[P] = S[P_A] + S[P_B]$. In fact, from Eq. (2.19):

$$S[P] = -\sum_{ij} P(A_i, B_j) \ln(P(A_i, B_j))$$

$$= -\sum_{ij} [P_A(A_i)P_B(B_j) \ln P_B(B_j) + P_B(B_i)P_A(A_j) \ln P_A(A_j)]$$

$$= -\sum_{i} [P_B(B_i) \ln P_B(B_i) + P_A(A_i) \ln P_A(A_i)] = S[P_B] + S[P_A]. \quad (2.21)$$

We see that the logarithm is important.

On the backward route from Shannon entropy to thermodynamic entropy, we shall see that the additivity of the entropy, is a somewhat reasonable requirement, in the case of weakly interacting system, exactly as additivity of the internal energy. Convexity lies in some sense at a deeper level, and is connected with the possibility of defining a concept of thermodynamic equilibrium. A question that has been considered over the years, is whether alternative definitions of entropies, "without logarithms", are possible, and could lead to a self-consistent thermodynamic theory of strongly interacting systems.

Coming back to the original definition of entropy Eq. (2.19), we may ask whether we could associate an entropy with a PDF, rather than with a probability distribution over a partition of Ω. The answer is trivially yes, once we indicate, with $\Delta x(x)$, the partition element, in which the value of the random variable x is located. From Eqs. (2.1) and (2.19):

$$S[P] = -\int dx\, \rho(x) \ln[\rho(x)\Delta x(x)],$$

which, in the case of a uniform partition $\Delta x(x) = \Delta x$, becomes

$$S[P] = -\int dx\, \rho(x) \ln \rho(x) - \ln \Delta x. \tag{2.22}$$

The entropy, in the case of a uniform partition, coincides therefore, to within a constant, with the "PDF entropy" $S_\rho = -\int dx\, \rho(x) \ln \rho(x)$. As we shall see in the following, the uniform partition is the natural choice, in most situations of interest for thermodynamics systems.

2.3 Sums of Random Variables: The Thermodynamic Limit

Quantities such as a random variable average, or the probability of an event, could be estimated as limits over a large number of trials of the corresponding sample average and frequencies. In fact, there exists an empirical interpretation of probability theory, in which probabilities are understood as limits of frequencies. The sample average of a random variable y is simply the arithmetic mean of the results of the measurement of the quantity y in a sequence of repeated identical experiments:

$$\langle y \rangle_N = \frac{1}{N} \sum_{k=1}^{N} y_k. \tag{2.23}$$

Similarly, the frequency of an event A can be expressed as the sample average of the indicator function, defined in Eq. (2.13):

$$\tilde{P}_N(A) = \langle \delta_A \rangle_N. \tag{2.24}$$

Of course, the sample average $\langle y \rangle_N$ is still a random variable, as its value will change, from a sequence of N experiments to the next. If the outcomes of different experiments can be considered as statistical independent, and if the PDF $\rho(y)$ has finite first and second moment, the **law of large number** will hold:

$$\langle \langle y \rangle_N \rangle = \mu_y, \qquad \sigma^2_{\langle y \rangle_N} = \frac{\sigma^2_y}{N}, \tag{2.25}$$

which expresses, formally, the notion that averages can be estimated as limits of sample averages. In the same hypotheses, the **central limit theorem** will hold, which tells us that the deviations $\langle y \rangle_N - \mu_y$ are normally distributed.

The link between statistics and thermodynamics lies in the fact that the macroscopic state of a system is described by variables that are, typically, sums of contributions by microscopic constituents:

$$\tilde{Y}_N = \sum_{k=1}^{N} y_k, \tag{2.26}$$

where N could be of the order of the Avogadro number $N_A \simeq 6 \cdot 10^{23}$ mol^{-1}. The quantity y_k may be e.g. the kinetic energy of the molecules in an ideal gas, and $\tilde{Y}_N$ the internal energy of the gas. The fact is that, in many instances (e.g. an ideal gas at thermodynamic equilibrium), the contributions from the microscopic constituents can be considered as independent identically distributed (i.i.d.) random variables. The macroscopic variable Y_N will then satisfy a relation in the form of the law of large numbers Eq. (2.25):

$$\mu_{\tilde{Y}_N} = N\mu_y; \qquad \sigma_{\tilde{Y}_N} = N^{1/2}\sigma_y. \tag{2.27}$$

Let us check this result. The relation $\mu_{Y_N} = N\mu_y$ descends immediately from linearity of the average operation. To obtain the second of Eq. (2.27), we write

$$\sigma^2_{\tilde{Y}_N} = \langle [\sum_{i=1}^{N}(y_i - \mu_y)]^2 \rangle = \sum_{ij=1}^{N} \langle (y_i - \mu_y)(y_j - \mu_y) \rangle, \tag{2.28}$$

and observe that, since the different y_i's are statistically independent, we must have, for $j \neq j$ $\langle (y_i - \mu_y)(y_j - \mu_y) \rangle = \langle (y_i - \mu_y) \rangle \langle (y_j - \mu_y) \rangle = 0$, and hence:

$$\sigma^2_{\tilde{Y}_N} = \sum_{i=1}^{N} \langle (y_i - \mu_y)^2 \rangle = N\sigma^2_y.$$

Equation (2.27) leads us to identify three relevant scales for the variable Y:

- A microscopic scale, fixed by μ_y and σ_y.
- A fluctuation scale $\sigma_Y = N^{1/2}\sigma_y$.
- A macroscopic scale, at which the condition of thermodynamic limit is considered statisfied. For $\mu_y \neq 0$, this macroscopic scale can be identified simply with $\mu_{Y_N} = N\mu_y$.

We see that, for $N \to \infty$ (and provided $\mu_y \neq 0$), fluctuations will become negligible on the scale of the mean $\mu_{\tilde{Y}_N} = N\mu_y$. The distribution $\rho(\tilde{Y}_N)$ will then becomes so narrow (on the scale of interest), that we can identify the instantaneous value of the variable $\tilde{Y}_N$, with the most probable value Y_N, that coincides in turn with the mean $\mu_{\tilde{Y}_N}$. This condition is commonly referred to, as a **thermodynamic limit** for the variable Y.

Note We can apply the concepts developed so far, to evaluate the density fluctuations in a gas in thermodynamic equilibrium. No external forces are present, so that the gas is spatially homogeneous. To fix the ideas, imagine that the gas is contained in a volume V, and indicate with N the total number of molecules. We choose, as macroscopic variable, the coarse grained density $\tilde{n}_{V_a}(\mathbf{x}_a, t) = \tilde{N}_a/V_a$, where $\tilde{N}_a$ is the instantaneous molecule count in the volume V_a centered around $\mathbf{x}_a$. Indicating with $\mathbf{x}_i(t)$ the instantaneous position of the ith molecule:

$$\tilde{n}_{V_a}(\mathbf{x}_a, t) = \frac{1}{V_a} \sum_{i=1}^{N} \delta_{V_a}(\mathbf{x}_i(t)).$$

From the property of the indicator function, Eq. (2.14): $\langle \delta_{V_a} \rangle = P(V_a)$, the probability to find the generic molecule in V_a. For a homogeneous gas, $P(V_a) = V_a/V$ and we find the obvious result, that the local density of the gas coincides with the mean density $\bar{n} = N/V$:

$$n_{V_a}(\mathbf{x}_a, t) = \bar{n}.$$

To calculate the fluctuation, we must determine $\sigma_{\delta V_a}^2$. We have

$$\langle \delta_{V_a}^2 \rangle = \langle \delta_{V_a} \rangle = V_a/V,$$

and therefore

$$\sigma_{\delta V_a}^2 = \langle \delta_{V_a}^2 \rangle - \langle \delta_{V_a} \rangle^2 = V_a/V(1 - V_a/V).$$

For V_a small, we can disregard the $(V_a/V)^2$, and we find for the fluctuation amplitude:

$$\sigma_{\tilde{N}_a}^2 = N\sigma_{\delta V_a}^2 \simeq N_a \Rightarrow \frac{\delta N_a}{N_a} = O(N_a^{-1/2}). \tag{2.29}$$

The fluctuation density, in a volume V_a, will be $\delta n_{V_a} = \delta N_a/V_a \sim (V_a \bar{n})^{-1/2} \bar{n}.\blacksquare$

It is worth stressing the physical content of the conditions of i.i.d. statistics, and existence of the first moments, μ_y and σ_y^2, in the derivation of Eq. (2.27). We see that these conditions express, in different ways, the need of a separation between microscopic and macroscopic scales.

Let us analyze first the condition of finiteness of the lowest order moments. As discussed in the appendix, if μ_y were infinite, the ratio $Y_N/N \equiv \langle y \rangle_N$ would grow indefinitely, as larger and larger values of y are picked up by the molecules in the sample. (Similar arguments would apply as regards σ_y^2 and $\sigma_{Y_N}^2$). Contrary to the finite μ_y case, in which all molecules contribute the same to Y_N, in the infinite μ_y case, only a few molecules would contribute to Y_N, with $y_k \sim Y_N$. The separation of scale between microscopic contributions and macroscopic world, would clearly be broken.

Let us pass to the condition of independence of the microscopic contributions. We see that this condition is essential in order for σ_{Y_N} to grow sublinearly in N and become negligible, for N large, compared with μ_{Y_N}. If, say, all molecules were equally correlated, we would have a quadratic σ_Y^2, and, in order to neglect fluctuations, we would have to be sure from the start that $\sigma_y \ll \mu_y$. The typical situation is that the molecules are correlated below a certain scale λ, while, at larger separations, they are basically independent. This would mean that, in place of a set of N independent molecules, we would have a bunch of independent molecule clumps, each of size $\sim\lambda$. In order for a thermodynamic limit argument, along the line of the one leading from Eq. (2.28) to Eq. (2.27), we need that the number of clumps is large. This requires, in turn, that λ is a microscopic quantity. We shall see in Sect. 3.3 that the condition of independence of the molecules at macroscopic scales, is basically a condition of fluctuations in the system being confined to microscopic scales.

The condition that the different y_i's are identically distributed, is just a different way of saying that the system must be macroscopic. An obvious condition for this is in fact that the spatial scale of variation of Y_N is itself macroscopic, which in turn guarantees the existence of portions of the system sufficiently large to be considered macroscopic, but not too much for the microscopic statistics inside, to be spatially inhomogeneous. An example better illustrates the situation. Think again Y_N as the density of a gas, coarse-grained at scale V_a (see note above), but relax the hypothesis of spatial homogeneity. It is clear that, to have a local meaning, the volume V_a must be smaller than the scale of variation for the density. At the same time, however, V_a must be large and the number of molecules in it to be large and fluctuations be negligible.

We conclude with the observation that there are systems that are better described by quantities that are not simple sums of microscopic contributions. An example is the interaction component of the internal energy, in a system of interacting particle. In the case of binary interactions: $U = \sum_{i>j} U_{ij}$, with U_{ij} the interaction energy of molecules i and j. In analogy with the problem of correlations, the situation can be brought back to that of an additive variable, in the form of Eq. (2.26), provided the interaction is short-ranged, meaning that the interaction length is microscopic. We would then have a bunch of microscopic clumps of molecules, such that the potential energy could be written as a sum over contributions: $U = \sum_i U_i^{clump}$, with U_i^{clump} the sum of the interaction energies of the pairs in the ith clump. Again, we need that the clump be microscopic, in order for a large number of them to be present, and a thermodynamic limit, along the lines of Eq. (2.27), to be possible.

2.4 Entropy and Thermodynamic Equilibrium

The most important characteristic of thermodynamic entropy, is growth in the approach to thermodynamic equilibrium, We could even arrive to define thermodynamic equilibrium, as that state of an isolated system in which entropy is maximum. After all, the statement that thermodynamic entropy is maximum, is perfectly equivalent to saying that quantities such as temperature, pressure, etc. are spatially uniform in the system.

It turns out that we can cook, out of the Shannon entropy, a perfectly reasonable thermodynamic entropy, that will itself be maximum at thermodynamic equilibrium.

Let us go back to the problem of a gas in a volume V, considered in the previous section. At thermodynamic equilibrium, and in the absence of external forces, the gas density will be uniform, and the probability to find a molecule, picked up at random, in a volume V_a, will be simply $P(V_a) = V_a/V$. We can partition V in volumes V_a, and calculate the Shannon entropy of the resulting distribution:

$$S[P] = - \sum_a P(V_a) \ln P(V_a). \tag{2.30}$$

We see immediately that, if the partition is uniform, the distribution $P(V_a)$, corresponding to thermodynamic equilibrium, will be the equiprobable distribution. This guarantees that the Shannon entropy is maximum at equilibrium, as required. We stress the probabilistic nature of this result. Nevertheless, since the probability $P(V_a)$ is associated with the mean density $\bar{n}$, $S[P]$ will describe the density profile that is actually observed in equilibrium in the gas. Substituting $P(V_a) = V_a/V$ into Eq. (2.30), we obtain

$$S[P] = \ln V/V_a. \tag{2.31}$$

Within constant factors, this is the right form of the spatial part of the entropy for an ideal gas (see Eq. (3.33)).

We could go one step further, and use the Shannon entropy to parameterize how improbable a certain configuration $\tilde{n}_{V_a}(\mathbf{x}, t)$ is, working with frequencies in place of probabilities:

$$S[\tilde{P}_N] = - \sum_a \tilde{P}_N(V_a) \ln \tilde{P}_N(V_a); \qquad \tilde{P}_N(V_a) = \tilde{N}_a/N. \tag{2.32}$$

This entropy is associated directly with the fluctuating densities $\tilde{n}_{V_a}$, as can be seen from the relation $\tilde{P}_N(V_a) = (\tilde{n}_{V_a}/\bar{n})P(V_a)$. We see that any fluctuation $\tilde{n}_{V_a} - \bar{n}$ is associated with a value of the instantaneous entropy $S[\tilde{P}]$ that is smaller than the equilibrium value $S[P]$. We stress the different origin of the differences $S[P] - S[\tilde{P}]$, and $S[P] - S[P^{NE}]$, with P^{NE} a generic non-equilibrium distribution. The first describes a microscopic effect (fluctuations); the second a macroscopic deviation from equilibrium of the system. This is revealed by the fact that the difference $P - \tilde{P}$ will go to zero in the thermodynamic limit, while $P^{NE} - P$, that represents a macroscopic condition (think e.g. of a configuration in which the gas is prepared in such a way that it occupies only a portion of V), will remain finite in the limit.

We need at this point to make an observation. In order to obtain a reasonable definition of entropy, the partition $\{V_a\}$ had to be uniform. We could have chosen, in principle, a different partition, in which certain regions of V were subdivided more finely than others. It is clear that a non-uniform partition would correspond to place different weight on information in different regions in V, with the most finely subdivided regions, the ones considered most important. The resulting entropy, however, would not have been maximum at equilibrium.

The important question is whether the uniform partition choice we have made, could have been guessed without knowledge of the character of the equilibrium state. After all, all points in the volume are equivalent for molecules interacting with short-range forces (once they are sufficiently away from the wall). Similarly, Galileian invariance suggests that uniform partition of velocity space should be the right choice, if we were to calculate the entropy associated with the velocity distribution of the molecules in the gas.

These statements can be made in fact more general: as it will be discussed in Sect. 5.2, the choice of a uniform partition has a lot to do with energy conservation, and with the geometry of the microscopic phase space of thermodynamic systems.

2.5 Stochastic Processes

The microscopic degrees of freedom of a thermodynamic system (think e.g. of the molecules in a gas) will be characterized, typically, by a random, unpredictable dynamics. Similar random behavior could be expected, at macroscopic scales, in non-equilibrium conditions. An example is turbulence. Quantities. such as the velocity of a molecule, or the fluid velocity in a turbulent could then be represented, as arrangement of random variables that depend on time (the velocity of the molecule), on space (a snapshot of the fluid velocity field in a turbulent flow) or even on space-time (the evolution of that turbulent flow). We introduce a minimum of terminology.

- We refer to a sequence of random variables, indexed in time, as a **stochastic process**. Indexing could be performed either on discrete or continuous time (discrete or continuous stochastic process).
- We speak of a **random field**, when the random variables areindexed in space or in space-time.

These definitions force us to reconsider concepts such as that of sample space, result of a measurement, and probability. To fix the ideas, let us focus, for the moment, on the case of a stochastic process, the generalization to the case of a random field being rather natural.

We notice at once that, while in the case of a random variable the result of an experiment was a single number, in the case of a stochastic process, the result is a "history" of the process, i.e., a function of time. In the case, say, of a noise generator, this could be the record of the signal during the time the generator was turned on.

- We shall call **realization** of the stochastic process, the particular function that is generated in the given experiment.

We see that, while in the case of a random variable, Ω was some real interval, in the case of a stochastic process, it will be a function space. Thus, while in the case of a random variable, the PDF $\rho(x)$ was a real function, in the case of stochastic process, the PDF will be a functional. In the case of a discrete stochastic process ψ, a realization would be a sequence $\{\psi(t_n), n = 1, 2, \ldots\}$, and the PDF would be essentially a function of a vector (possibly with infinite components) $\rho(\{\psi(t_n), n = 1, 2, \ldots\})$. In the case of a continuous stochastic process, the argument of ρ would be a function of real variable $\psi = \psi(t)$. We shall use the square bracket notation $\rho[\psi]$ to indicate simultaneous dependence of ρ on all values of $\psi(t)$ in the domain of t. An example:

$$\rho[\psi] = \mathcal{N} \exp \left(- \int_0^T |\psi(t)|^2 dt \right). \tag{2.33}$$

It is difficult, in general, to work with objects such as Eq. (2.33). In most situations, we must content ourselves with a reduced description, such as the one provided e.g. by the joint PDF

$$\rho(\{\psi_i, t_i\}) \equiv \rho(\{\psi(t_i) = \psi_i, i = 1, 2, \ldots\}), \tag{2.34}$$

where the t_i's represent generic sampling times of the continuous stochastic process.

A basic question we may want to ask is the evolution of a stochastic process, given a certain past history. We may want to consider, therefore, also conditional PDF's such as

$$\rho(\{\psi_i, t_i\}|\{\bar{\psi}_j, \tau_j\}) \equiv \rho(\{\psi(t_i) = \psi_i\}|\{\psi(\tau_j) = \bar{\psi}_j\}), \tag{2.35}$$

where $t_i > \tau_j \ \forall i, j$. More in general, we may consider the PDF resulting from an ensemble of trajectories stemming from initial conditions distributed with a generic PDF $\rho_0(\{\bar{\psi}_j, \tau_j\})$:

$$\rho(\{\psi_i, t_i\}) = \int \prod_j d\bar{\psi}_j \rho(\{\psi_i, t_i\}|\{\bar{\psi}_j, \tau_j\})\rho_0(\{\bar{\psi}_j, \tau_j\}) \tag{2.36}$$

We stress that the PDF $\tilde{\rho}_0$, that acts as initial condition for ρ, is arbitrary. This seems to suggest that, in order to define a stochastic process, we should impose always some kind of "initial condition". That is to say, the PDF in Eq. (2.34) will contain in general information on the way the process is turned on. The problem of isolating the intrinsic properties of the stochastic process can be solved if memory of the initial conditions decays with time:

$$\lim_{\tau_j \to -\infty} \rho(\{\psi_i, t_i\}|\{\bar{\psi}_j, \tau_j\}) = \rho(\{\psi_i, t_i\}). \tag{2.37}$$

This is a condition of statistical independence of the values of the stochastic process at large time separations, that is actually rather difficult not to be satisfied. We have basically two possibilities:

- The system is deterministic. An example is the "ping-pong", a discrete time process that oscillates indefinitely between two states 0 and 1. Thus, even for $t \to \infty$: $P(\psi, t|\psi_0, 0) = \delta_{\psi \psi_0}$, if t is even, while $P(\psi, t|\psi_0, 0) = 1 - \delta_{\psi \psi_0}$, if t is odd.
- The ambient space of the process is decomposable in subdomains that are mutually inaccessible, in the sense that, if ψ was initially in one of them, it will remain there forever. An example of such a situation, is a molecule that moves in a volume divided in two parts by an impermeable wall.

A stochastic process will be said to be in a condition of **stationary statistics**, if the PDF's in Eq. (2.34) are invariant by time translations:

$$\rho(\{\psi_i, t_i + T\}) = \rho(\{\psi_i, t_i\}), \quad \forall T. \tag{2.38}$$

This means that the PDF's $\rho(\{\psi_i, t_i\})$ depend solely on the time differences $t_i - t_j$, and that the PDF $\rho(\psi, t) = \rho(\psi)$ is itself independent of the time t. If the limit in the LHS of Eq. (2.37) exists (this basically rules out deterministic systems such as the ping-pong), one speaks of **equilibrium statistics**. The limit

$$\rho(\{\psi_i, t_i\}|\bar{\psi}) \equiv \rho(\{\psi_i, t_i\}|\bar{\psi}, \tau \to -\infty\})$$

is called an **equilibrium PDF**. (Notice that a single initial condition $\bar{\psi}$ is sufficient to identify the portion of ambient space where a given realization evolves). If Eq. (2.37) is satisfied, which implies that the stochastic process ambient space is undecomposable, the equilibrium statistics will also be unique. This means that any initial condition $\rho_0(\psi, t_0)$ will evolve, after a sufficiently long time, into the single equilibrium PDF $\rho(\psi)$. If the ambient space were decomposable, the same statement would remain true, but only locally in each of the subdomains in which the ambient space is subdivided. If ρ_0 had support in just one of such subdomains, which can be identified by one of its points $\bar{\psi}$: $\rho_0(\psi, t_0) \longrightarrow \rho(\psi|\bar{\psi})$. Notice finally, that a PDF could be stationary, and, at the same time, not be an equilibrium PDF. An example is again provided by the ping-pong: the PDF $\rho(1) = \rho(0) = 1/2$ is obviously stationary, but it is not true that, after a long enough time, any initial PDF would collapse to the stationary PDF.

2.5.1 The Ergodic Property

Exactly as moments can be used to describe the statistics of random variables, **correlation functions** in the form

$$\langle \psi(t_1)\psi(t_2)\ldots\rangle, \tag{2.39}$$

can be used to describe the statistics of a stochastic process $\psi(t)$. The two-time (subtracted) correlation function

$$C(t, t') = \langle \psi(t)\psi(t')\rangle - \langle \psi(t)\rangle\langle \psi(t')\rangle, \tag{2.40}$$

in particular, allows to estimate the degree of statistical independence of the values at different times of the stochastic process. We have in fact, by definition:

$$\langle \psi(t)\psi(t')\rangle = \int d\psi \int d\psi'\, \rho(\psi, t; \psi', t'),$$

and, in the case of independent values of ψ, i.e. if $\rho(\psi, t; \psi', t') = \rho(\psi, t)\rho(\psi', t')$:

$$\langle \psi(t)\psi(t')\rangle = \langle \psi(t)\rangle\langle \psi(t')\rangle.$$

We can define a **correlation time**, that tells us at what time separation, correlations can be considered negligible. The definition commonly adopted is

$$\tau_\psi(t) = \frac{1}{C(t, t)} \int\limits_0^\infty C(t, t + \tau)d\tau \tag{2.41}$$

(as it is to say, area under $C(t, t + \tau)$ equal to basis – i.e. $\tau_\psi(t)$ – times height – i.e. $C(t, t)$). In terms of a single realization of the stochastic process, $\tau_\psi(t)$ tells us how long we must wait until $\psi(t + \tau)$ differs much from $\psi(t)$. In other words, the correlation time gives us the time scale of variation of the process.

The averages that appear in the equations that have been written up to now, could be estimated by means of sample averages. We wonder whether, in the case of a time-independent process, such averages could be replaced by time averages in the form

$$\langle \psi \rangle_T = \frac{1}{T} \int\limits_0^T \psi(t)dt. \tag{2.42}$$

In other words, whether, instead of estimating averages out of sample averages over multiple realization of the same process, we could perform this operation, working with only one realization of that process.

An example, in which the sample and time averages are trivially equivalent, is that of the stochastic process, obtained putting in sequence the independent measurements of a random variable y. It is rather clear that carrying out a sample average using separate realizations of such process, would be pointlessly costly.

- The condition that averages $\langle . \rangle$ could be approximated, for long enough sample times, by time averages, is called an **ergodic property** for the process.

The first condition for an ergodic property is clearly that the statistics of the process be stationary on the sampling time of the average.

The second condition is that Eq. (2.37) be satisfied. This implies in particular that the ambient space of the stochastic process is indecomposable. If this condition were not satisfied, any PDF in the form, say,

$$\rho(\psi) = \int d\bar{\psi}\, \rho_0(\bar{\psi})\rho(\psi, t|\bar{\psi}, \tau \to -\infty)$$

could be an equilibrium PDF for the process, and time averages starting from the subdomain of ψ, containing $\bar{\psi}$, could estimate averages only for $\rho(\psi, t|\bar{\psi}, \tau)$. More in general, Eq. (2.37) implies statistical independence of ψ at infinite time separation. This guarantees that we can imagine the sampling interval T as a sequence of many intervals, such that the values of the quantity, in any two of them, are approximately independent. Actually, in order for ergodicity to be satisfied, it is enough that the correlations of the quantity to be averaged decay to zero for $t \to \infty$.

The third condition, that parallels the one in the law of large numbers, is that average and variance of the quantity in exam are finite.

We illustrate the result, in the case the quantity to average is simply ψ, and the correlation time τ_ψ is finite. We proceed as in Sect. 2.3, comparing $\langle\psi\rangle$ with $\langle\psi\rangle_T$, and calculating $\sigma^2_{\langle\psi\rangle_T}$. Exchanging average and time integral in $\langle\langle\psi\rangle_T\rangle$, we obtain at once

$$\langle\langle\psi\rangle_T\rangle = \langle\langle\psi\rangle\rangle_T = \langle\psi\rangle.$$

Let us indicate, by $\tilde{\psi}(t) = \psi(t) - \langle\psi\rangle$, the deviation from the mean, and write

$$\sigma^2_{\langle\psi\rangle_T} = \frac{1}{T^2}\int_0^T dt \int_0^T dt'\, \langle\tilde{\psi}(t)\tilde{\psi}(t')\rangle = \frac{1}{T^2}\int_0^T dt \int_0^T dt'\, C(t - t').$$

Now, for $T \gg \tau_\psi$, we can approximate the integral in dt', for almost all values of t, with an integral between $\pm\infty$ (at $t' = 0$ and at $t' = T$, if $|t - t'| \gg \tau_\psi$, $C(t - t')$ will be in any case already zero). Exploiting Eq. (2.41), we find:

$$\sigma^2_{\langle\psi\rangle_T} \approx \frac{1}{T^2}\int_0^T dt \int_{-\infty}^{+\infty} d\tau\, C(\tau) = \frac{2\tau_\psi}{T}C(0) \equiv \frac{2\tau_\psi}{T}\sigma^2_\psi. \tag{2.43}$$

The variance $\sigma^2_{\langle\psi\rangle_T}$ decreases with the sampling time T, that plays here the same role as the sample size N in the law of large numbers Eq. (2.25).

The line of reasoning that we have followed so far, could be adapted to the random field case, introducing space correlations in the form

$$C(\mathbf{x}, \mathbf{y}) = \langle\psi(\mathbf{x})\psi(\mathbf{y})\rangle - \langle\psi(\mathbf{x})\rangle\langle\psi(\mathbf{y})\rangle.$$

The concept of statistical stationarity translates in that of spatially uniform statistics, as a requirement of independence on spatial translation. This means, in particular

$C(\mathbf{x}, \mathbf{y}) \rightarrow C(\mathbf{x} - \mathbf{y})$. In many instances, the statistics is also isotropic, meaning that $C(\mathbf{r}) \rightarrow C(r)$. In this case, definition of a orientation-independent **correlation length** is possible:

$$\lambda_\psi = \left(\frac{1}{C(0)} \int_0^\infty d^3 r \, C(r) \right)^{1/3} .$$

In the case of spatially uniform statistics, we can define the concept of spatial average; for instance, in the case of ψ:

$$\langle \psi(\mathbf{x}) \rangle_V = \frac{1}{V} \int_V d^3 y \, \psi(\mathbf{y}),$$

where V is centered around $\mathbf{x}$. An ergodic property will be satisfied in space, if the statistics is uniform on the sampling volume and λ_ψ is small. A simple generalization of the calculation leading to Eq. (2.43), gives in fact

$$\sigma^2_{\langle\psi\rangle_V} \approx \frac{\lambda_\psi^3 \sigma_\psi^2}{V} .$$

This result brings us back to the considerations in Sect. 2.3, on the necessity that correlations between molecules decay at microscopic scales, in order for a thermodynamic limit to be satisfied. The molecule count in a volume, in that case, could be seen as an integral over a density in that same volume. Correlations among molecules, would translate into correlation among density fluctuations, and the condition that correlations among molecules, occur only at microscopic scales, corresponds to the condition that λ_n itself be microscopic.

2.5.2 Markov Processes

We have dealt so far with the "kinematics" of a stochastic process. We would like to pass to the "dynamics", i.e. to the task of writing evolution equations for time-dependent probabilities and PDF's. The simplest object that we have considered, is the conditional PDF $\rho(\psi, t | \bar\psi, \tau)$.

We concentrate, for the moment, on the case of stochastic processes that take discrete values $\psi_n = n\Delta\psi$ (called **states** of the system), at discrete times $t_k = k\Delta t$. Given a stochastic process $\psi(t_k)$, we utilize the notation

$$P(\{n_i; i = 0, 1, \ldots\}) \equiv P(\{\psi(t_i) = \psi_{n_i}, i = 0, 1, \ldots\}).$$

Similar notations could be introduced in the case of conditional probabilities. We are interested, in particular, in the quantity $P(n_{i+1} | \{n_k; k = 0, 1, \ldots, i\})$, that is connected with the probability of one realization (starting at discrete time $k = 0$), by the standard relation:

$$P(n_{i+1}|\{n_k; k = 0, 1, \ldots, i\}) = \frac{P(\{n_k; k = 0, 1, \ldots, i+1\})}{P(\{n_k; k = 0, 1, \ldots, i\})}.$$

We shall consider only two key cases:

- Totally uncorrelated processes (e.g. the process resulting from repeated independent measurements of a random variable):

$$P(n_{i+1}|\{n_k; k = 0, 1, \ldots, i\}) = P(n_{i+1}).$$

- Systems in which, knowledge of the state of the system at a certain time, makes all information on the previous states of the system, irrelevant (example: the random walker that decides each new step with a throw of dice. Its position at any given step will depend only on its position one step earlier, and on the result of the last throw of dice). In formulae:

$$P(n_{i+1}|\{n_k; k = 0, 1, \ldots, i\}) = P(n_{i+1}|n_i). \tag{2.44}$$

A process with such characteristics, is called a **Markov process**.

We see that it is possible to derive local evolution equations for Markov processes, based solely on knowledge of the **transition probability** $P(n_{i+1}|n_i)$. The role of building block played by the transition probability, is particularly apparent in the case of the random walker, as $P(n_{i+1}|n_i)$ tells us precisely with which probability the walker goes, in one step, from location n_i to n_{i+1}.

The starting point is the following obvious relation, which descends from Eq. (2.44):

$$P(n_{i+1}, n_i, n_0) = P(n_{i+1}|n_i, n_0)P(n_i|n_0)P(n_0) = P(n_{i+1}|n_i)P(n_i|n_0)P(n_0).$$

From here we get

$$P(n_{i+1}, n_i|n_0) = P(n_{i+1}|n_i)P(n_i|n_0).$$

Summing over the intermediate states, we obtain the following relation, called the **Chapman-Kolmogorov** equation:

$$P(n_{i+1}|n_0) = \sum_{n_i} P(n_{i+1}|n_i)P(n_i|n_0). \tag{2.45}$$

In plain words: the probability of finding the system at a given time in a certain state, is obtained, summing on the probabilities of the possible states in which the system could lie at the previous time, times the transition probability from the past to the current state.

The starting point in the derivation of the Chapman-Kolmogorov equation (2.45), was $P(n_{i+1}, n_i, n_0)$. We could have used a generic probability $P(n_k, n_i, n_0)$, with $k > i > 0$ generic, and, proceeding in the same way, we would have obtained

$$P(n_k|n_0) = \sum_{n_i} P(n_k|n_i)P(n_i|n_0).$$

Similarly, we can multiply both sides of the Chapman-Kolmogorov equation by an arbitrary probability $P(n_0)$, sum over the initial state n_0, and obtain the evolution equation for the time dependent probability

$$P(n_{i+1}) = \sum_{n_0} P(n_i|n_i)P(n_i), \qquad (2.46)$$

that has solution, once the initial condition $P(n_0)$ is given.

The most complete statistical description we could reach, of a discrete stochastic process such as the one considered, is provided by the joint probability $P(\{n_i, i = 1, 2, \ldots\}|n_0)$. Knowledge of the transition probability $P(n_{i+1}|n_i)$ is again enough to obtain such a description. By repeated application of the relation

$$P(\{n_i, i = 1, 2, \ldots, k+1|n_0\})$$
$$= P(\{n_{k+1}|\{n_i, i = 0, 2, \ldots, k\})P(\{n_i, i = 1, 2, \ldots, k\}|n_0)$$
$$= P(\{n_{k+1}|n_k)P(\{n_i, i = 1, 2, \ldots, k\}|n_0),$$

we obtain the following so-called **Markov chain expression**:

$$P(n_i, n_{i-1}, \ldots, n_1|n_0) = P(n_i|n_{i-1})P(n_{i-1}|n_{i-2})\ldots P(n_1|n_0). \qquad (2.47)$$

If the transition probability is independent of time,

$$P(n_{i+1} = m|n_i = q) \equiv P(q \rightarrow m)$$

(in which case, the Markov process is said to be itself time-independent), and $P(n_0)$ is a stationary solution of the Chapman-Kolmogorov equation, multiplication of Eq. (2.47) by $P(n_k)$ will lead to an expression for $P(\{n_i, i = 0, 1, \ldots\})$ that is invariant by time translation, i.e. to stationary statistics.

2.6 Coarse-Graining and Continuous Limit

The passage from a microscopic description of a physical system to a macroscopic one, can be envisioned as a coarse-graining operation, in which microscopic degrees of freedom of the system are averaged away. Thermodynamic variables such as pressure, temperature and density can then be seen as local (volume) averages of microscopic quantities.

By and large, the coarse-graining process can be visualized as a sequence of three steps:

- Identification of a **microscale** of the system, below which the details of the dynamics are unknown (and supposed irrelevant). In the case of a gas, this could be the scale of the individual molecules.
- Introduction of a **mesoscale** l, where the dynamics is "less complicated" than at the microscale. This could mean many things. In the case of the gas, again, it could be the scale at which a thermodynamic limit begins to be possible (i.e. fluctuations are small). In this case, the mesoscale is precisely the smallest scale at which a description of the system in terms of macroscopic variables (e.g. the density) becomes possible.
- Identification of the **macroscale** L of the system, as the scale of variation for the macroscopic variables.

In the example of the gas, considered in Sect. 2.3, the microscopic description was provided by knowledge of the exact position of all molecules, which could be expressed through the fine-grained expression for the density field

$$\tilde{n}(\mathbf{x}, t) = \sum_{i=1}^{N} \delta(\mathbf{x} - \mathbf{x}_i(t)), \tag{2.48}$$

with $\mathbf{x}_i(t)$ the position of the ith molecule. The mesoscopic description of the same system corresponds to coarse-graining the density field $\tilde{n}(\mathbf{x}, t)$ at the scale l identified by the volume V_a:

$$\tilde{n}_{V_a}(\mathbf{x}_a, t) = \langle \tilde{n}(\mathbf{x}_a, t) \rangle_{V_a} \equiv N_a(\mathbf{x}_a, t)/V_a.$$

The important point is that, if the coarse graining scale is much larger than that of the microscopic constituents, the coarse-grained quantity will take a (more or less) continuous character. This is a **continuous limit** for the coarse-grained quantity, that corresponds to our notion of macroscopic variables as continuous fields at the macroscopic scales of interest for the system.

Similar coarse-graining procedures can be carried out with respect to time, if, in place of a random field, we consider a stochastic process.

Once coarse graining has been carried out, and a continuous limit has been taken, it may become necessary to discretize again the variables, e.g. to solve numerically the evolution equations. In certain sense, the operation is akin to follow backward the coarse graining procedure, passing from a mesoscopic description of the system to a fictitious microscopic one. The fact that different microscopic dynamics lead to the same macroscopic behaviors is interesting, as it suggests us that only a small subset of features at the microscale could actually be relevant at the macroscopic level. It could then be expedient to seek from the start a model of the microscopic dynamics that is as simple as possible.

Unfortunately, the process of building a discrete model that has an appropriate continuous limit is far from trivial. In fact, it is not guaranteed at all that a microscopic dynamics will admit in general a continuous limit. This is why we stressed earlier the fact that continuity had to be understood in a "more or less" sense. There is a general problem of lack of regularity in the realizations of continuous stochastic processes and random fields, that, as we shall see in the next section, disappears only if one decides to deal with average quantities. An example of such lack of regularity is provided already in the case of the random walker.

Indicate with Δt and Δx the time constant and the step length of the walker (both assumed constant and position independent). Focus on the case of an unbiased, spatially uniform and stationary random walk process, such that the jump probability can be written in the form

$$P(x \rightarrow x + \Delta x) = P(x \rightarrow x - \Delta x) = P_J, \qquad P(x \rightarrow x) = 1 - 2P_J, \qquad (2.49)$$

with P_J a constant. We want to take the continuous limit of this process by considering macroscopic times t and separations $x(t) - x(0)$, such that $t \gg \Delta t$ and $x(t) - x(0) \gg \Delta x$. From Eq. (2.49), we obtain immediately:

$$\langle |x(t) - x(0)|^2 \rangle = 2P_J \sum_{i=1}^{t/\Delta t} \Delta x^2 = 2 \frac{P_J \Delta x^2}{\Delta t} t. \qquad (2.50)$$

The two quantities Δx and Δt are seen as zero at the macroscopic scales fixed by $x(t) - x(0)$ and t. In order for Eq. (2.50) to make sense, however, it is necessary that the quantity $\kappa = P_J \Delta x^2 / \Delta t$, remains finite as Δt and Δx are sent to zero at fixed P_J. In other words, in order for a continuous limit to exist, it is necessary that $\Delta x = O(\Delta t^{1/2})$. We can verify that the realizations of the continuous random walk process obtained, taking this limit, are continuous. In fact, Eq. (2.50) tells us that the typical displacement $|x(t) - x(0)|$ goes to zero as $t^{1/2}$, so that the typical realization of the process is continuous.

But they are not differentiable.

If we try to define a velocity scale as limit of a finite difference ratio, we obtain in fact an infinite result:

$$\dot{x} \sim \lim_{t \to 0} \frac{\langle |x(t) - x(0)|^2 \rangle^{1/2}}{t} \propto \lim_{t \to 0} t^{-1/2} = \infty.$$

As we shall in the next section, in order to find a well-behaved quantity in the limit considered, we must consider average quantities such as the PDF $\rho(x, t)$. In alternative, coarse graining of the stochastic process must be carried out:

$$x(t) \rightarrow x_\tau(t) = \int dt' G_\tau(t - t') x(t'), \qquad (2.51)$$

where G has support $\sim \tau$, with $\int dt\, G_\tau(t) = 1$ and $G_\tau(t) > 0$. The regularity properties of $\dot{x}_\tau(t)$ would be therefore those of G_τ. (The simplest choice $G_\tau(t) = (1/(2\tau)) \delta_{[-\tau,\tau]}(t)$, would be good to cure discontinuities in $\dot{x}$, but would not be enough to guarantee differentiability of $\dot{x}_\tau$).

Note We provide an example of the way in which coarse-graining can be used to simplify a complicated microscopic dynamics. Consider a discrete random walker, whose steps have variable length and are correlated in time. Due to time correlations, the walker dynamics will be non-Markovian. Imagine for simplicity that correlations exist only between successive steps. Indicating by Δn_i the step at discrete times $t_i = i\Delta t$:

$$\langle \Delta n_i \Delta n_{i\pm 1}\rangle = c\sigma_{\Delta n}^2, \quad 0 < c < 1; \quad \langle \Delta n_i \Delta n_{i\pm k}\rangle = 0, \quad k \geq 2.$$

Let us coarse-grain time at a scale $\overline{\Delta t}$ that contains many microscale intervals Δt. We shall have

$$\overline{\Delta n_i} = \sum_{j=1}^{m} \Delta n_{i+j},$$

from which, exploiting $c > 0$:

$$\sigma_{\overline{\Delta n}}^2 \geq m\sigma_{\Delta n}^2,$$

while

$$\langle \overline{\Delta n_i}\, \overline{\Delta n_{i\pm 1}}\rangle = \langle \Delta n_i \Delta n_{i\pm 1}\rangle = c\sigma_{\Delta n}^2$$

(only the last contribution Δn_i in an interval $\overline{\Delta t}$, and the first one in the next interval, are correlated). We thus see that the correlation strength, parameterized by the ratio $\langle \overline{\Delta n_i}\, \overline{\Delta n_{i\pm 1}}\rangle / \sigma_{\overline{\Delta n}}^2$ becomes negligible for large m. In other words, coarse-graining in time brings us back to an independent step situation, such that n_i becomes a Markov process.∎

2.7 Master Equation and Its Applications

As we have said, the regularity problems in the continuous limit of a Markov process, will disappear, if we work with averages rather than with realizations. In particular, we can carry out a continuous limit on both arguments of the probability $P(n, k) \equiv P(\psi(t_k)=\psi_n)$; $\psi_n = n\Delta\psi$; $t_k = k\Delta t$, and be able to write evolution equations for such probability (and associated PDF's) in differential, or integro-differential form.

Our starting point is the Chapman-Kolmogorov equation (2.45), that we can rewrite in the form

$$P(m, i+1) - P(m, i) = \sum_k P(k \to m)P(k, i) - P(m, i)$$

$$= \sum_{k \neq m} [P(k \to m)P(k, i) - P(m \to k)P(m, i)], \quad (2.52)$$

where we have utilized the relation $\sum_k P(m \to k) = 1$, and the step $\sum_k \to \sum_{k \neq m}$ is justified by the fact that the term $k = m$, in the two addends in square brackets, cancel one another.

If the difference $P(m, i + 1) - P(m, i)$ is small on the scale of $P(m, i)$, which is equivalent to say that the transition probability is small: $P(k \to m) \ll 1$, it is meaningful to introduce the transition rate (transition probability per unit time):

$$W(k \to m) = \frac{P(k \to m)}{\Delta t}.$$

The transition probability, in a time interval $\overline{\Delta t} > \Delta t$, will thus be

$$P(m, t + \overline{\Delta t}|k, t) = W(k \to m)\overline{\Delta t} + O((W\overline{\Delta t})^2),$$

where the first term to RHS is the probability of just one transition in the interval, while the second is the contribution from multiple transitions. The continuous limit corresponds to consider the coarse graining scale $\overline{\Delta t}$ much larger than Δt, but, at the same time, small enough for $O((W\overline{\Delta t})^2)$ contributions to $P(m, t + \overline{\Delta t}|k, t)$ to be negligible. Equation (2.52) can be cast therefore in the differential form

$$\frac{\partial P(m, t)}{\partial t} = \sum_{k \neq m} [W(k \to m)P(k, t) - W(m \to k)P(m, t)]. \qquad (2.53)$$

This is called the **master equation** for the process. The Markov process, described in Eq. (2.53), is continuous in time, but is still defined in a discrete state space. The condition to take the continuous limit also in ψ, is that the variations of $P(m, t)$, this time with respect to m, be small on the scale of $P(m, t)$. We can in this case introduce the PDF $\rho(\psi_m, t) = P(m, t)/\Delta\psi$, and the transition rate density

$$w(\psi_k \to \psi_m) = \frac{W(k \to m)}{\Delta\psi}.$$

The master equation will take the integro-differential form:

$$\frac{\partial \rho(\psi, t)}{\partial t} = \int d\psi' \, [w(\psi' \to \psi)\rho(\psi', t) - w(\psi \to \psi')\rho(\psi, t)]. \qquad (2.54)$$

The non-local character of the above equation, reflects the fact that the transitions at the microscopic timescale Δt, occur at a macroscopic scale. In order for a clear separation to exist between macroscopic world and microscopic stochastic dynamics, it is necessary that the transitions occur at the microscopic scale fixed by $\Delta\psi$. We shall provide below, two important examples of systems with such clear separation of scales, whose evolution will be described by partial differential equations, rather than integro-differential equations for the probability.

2.7.1 The Random Walk

We have already seen examples of this process in the previous sections: an individual that at discrete times $t_k = k\Delta t$, with equal probabilities P_J, may jump either left or right one discrete step Δx, and with probability $1 - 2P_J$, may remain where it is. The states of the stochastic process are therefore the discrete positions $x_n = n\Delta x$.

We have seen in Sect. 2.6, that a continuous limit requires that the quantity $\kappa = P_J(\Delta x)^2 \Delta t$ remains finite for $\Delta x, \Delta t \to 0$, which imposes $\Delta x = O((\Delta t)^{1/2})$. If we plug into the master equation (2.52), we find

$$\frac{P(m, i+1) - P(m, i)}{\Delta t} = \kappa \frac{P(m+1, i) + P(m-1, i) - 2P(m, i)}{(\Delta x)^2}. \tag{2.55}$$

We recognize immediately, to RHS, the finite difference representation of $\partial^2 P/\partial x^2$. Taking the continuous limit, and introducing the PDF $\rho(x_n, t_k) = P(n, k)/\Delta x$, we obtain the **diffusion equation** (or **heat equation**):

$$\frac{\partial \rho(x, t)}{\partial t} = \kappa \frac{\partial^2 \rho(x, t)}{\partial x^2}, \tag{2.56}$$

in which κ takes the name of **diffusivity** of the random walker. The dynamics described by Eq. (2.56) corresponds to the progressive spreading of the initial walker distribution. The nature of the equation suggests the existence of similarity solutions in the form

$$G(x, t) = g(x^2/t).$$

Substituting into Eq. (2.56), we find in fact the Gaussian solution

$$G(x, t) = \frac{1}{\sqrt{4\pi\kappa t}} \exp\left(-\frac{x^2}{4\kappa t}\right), \tag{2.57}$$

corresponding to the initial condition $G(x, 0) = \delta(x)$. The solution $G(x, t)$ is the Green function for the problem in an infinite domain. The evolution of the PDF ρ, out of the generic initial condition $\rho(x, 0)$, will be therefore

$$\rho(x, t) = \int dx' \, G(x - x', t)\rho(x, 0). \tag{2.58}$$

If the domain is finite, determination of Green function of the problem will be more complicated, but the stationary solutions to Eq. (2.56) are easy to find. In the case of a finite domain with impermeable walls, $A = [x_1, x_2]$, in particular, we find the spatially uniform solution $\rho(x) = 1/(x_2 - x_1)$. Notice that this was the maximum-entropy equilibrium distribution, described in Sect. 2.4. In fact, it is possible to prove that the dynamics described by Eq. (2.56) corresponds always to growth of the PDF entropy of $\rho(x, t)$ (see problem 4 at the end of this chapter).

Let us go back to analysis of the diffusion equation (2.56). We see that its structure is that of a continuity equation

$$\frac{\partial \rho(x, t)}{\partial t} + \frac{\partial J(x, t)}{\partial x} = 0, \tag{2.59}$$

in which $J = -\kappa \, \partial \rho / \partial x$ has the meaning of a probability flux. We can associate the current J with a **current velocity**

$$u(x, t) = J(x, t)/\rho(x, t) = -\kappa \frac{\partial \ln \rho(x, t)}{\partial x},$$

that is the average velocity of the walkers that cross coordinate x at time t. It must be stressed that this quantity does not represent an average velocity of the individual walkers (this average is indeed zero), rather, it is the cumulative result of the fact that, if there are more walkers at one side of x, more walkers will cross x from that side. The current J has the form of a **flux-gradient** relation, $J = -\kappa \, \partial \rho / \partial x$, whose effect is to smooth out the inhomogeneities in ρ.

We can use Eq. (2.59) to write the evolution equation for the probability $P_A(t) = \int_A dx \, \rho(x, t)$, of finding the walker in a domain $A = [x_1, x_2]$, in the form

$$\dot{P}_A(t) = J(x_1, t) - J(x_2, t),$$

(the boundary conditions, appropriate to describe a domain with impermeable walls, will thus be the zero-gradient condition $\partial_x \rho(x, t)|_{x=x_{1,2}} = 0$).

In general, evolution equations for the PDF of stochastic processes cannot be solved in closed form. A partial solution of such equation, can nevertheless be obtained, in some cases, in terms of moments.

We can convert an evolution equation for a PDF into one for the moments $\langle x^n \rangle$, by multiplying both sides of the equation, by the desired power of x, and integrating over x. Let us carry out the procedure explicitly in the case of Eq. (2.56). Indicate by $f(x)$ the quantity, of whose average we want to determine the evolution. Multiplying the LHS of Eq. (2.56) by $f(x)$, and integrating over x, we find:

$$\int dx \, f(x) \frac{\partial \rho(x, t)}{\partial t} = \frac{\partial}{\partial t} \int dx \, f(x) \rho(x, t) = \frac{d}{dt} \langle f(x(t)) \rangle.$$

Doing the same thing with the RHS of that same equation, and eliminating derivatives on ρ, by repeated integration by parts, we obtain

$$\int dx \, f(x) \frac{\partial^2 \rho(x, t)}{\partial x^2} = \left[f(x) \frac{\partial \rho(x, t)}{\partial x} \right]_a^b - \left[\frac{\partial f(x)}{\partial x} \rho(x, t) \right]_a^b + \left\langle \frac{\partial^2 f}{\partial x^2} \right\rangle,$$

where a and b are the extrema of the walker domain. If the domain is infinite, the boundary terms can be eliminated,[1] and we are left with an equation in the form

$$\frac{d}{dt}\langle f \rangle = \kappa \langle f'' \rangle.$$

Taking $f(x) = x$, we find $x(t) = x(0)$, and taking $f(x) = x^2$, we obtain once more Eq. (2.50).

2.7.2 Extinction Process

In this case, the stochastic process describes the evolution of the number N of individuals in a sterile community. We count the individuals at discrete times $t_k = k\Delta t$ so closely spaced, that it is very unlikely that more than one individual dies in the interval Δt. We suppose that age does not affect the death process, as it would occur if the individuals were nuclei undergoing radioactive decay. In the same way, we suppose the death events to be independent, so that the probability to observe a death in the population in the interval Δt, will be N times the probability of death of one individual in the same time interval. If we indicate by Γ_D the death probability per unit time of one individual, the transition probability for the system will be

$$P(N \to N - 1) = W(N \to N - 1)\Delta t = N\Gamma_D \Delta t.$$

No other transitions are possible. The master equation for the process will be therefore

$$\frac{\partial P(N, t)}{\partial t} = \Gamma_D[(N + 1)P(N + 1, t) - NP(N, t)] \simeq \Gamma_D \frac{\partial (NP(N, t))}{\partial N},$$

which could be converted into an equation for the PDF $\rho(N, t) = P(N, t)/\Delta N \equiv P(N, t)/1 = P(N, t)$:

$$\frac{\partial \rho(N, t)}{\partial t} = \Gamma_D \frac{\partial (N\rho(N, t))}{\partial N}. \tag{2.60}$$

Proceeding as in the random walk case, we can convert the evolution equation for the probability, into an evolution equation for the mean population and its fluctuation.

As in the case of the diffusion equation (2.56), the evolution equation (2.60) can be cast in the form of a continuity equation, such as (2.59), with probability flux $J(N, t) = -N\Gamma_D \rho(N, t)$. In this case, the current velocity $u(N, t) = -N\Gamma_D$ is the speed at which a community of N individuals shrinks, due to the death

---

[1] The simpler way to be convinced of this fact is that the discrete walker will not be able, at finite t, to reach infinity; thus $\rho = 0$ there. The principle extends to generic discrete Markov processes with local transitions.

process. The end result is extinction, corresponding to an equilibrium distribution $P(N, t \to \infty) = \delta_{N,0}$. This is to be contrasted with the maximally spread-out state, towards which tends the random walker distribution. We thus see that a random microscopic dynamics may equally well lead to a statistical equilibrium state, whose Shannon entropy is maximum (the random walker), or zero (the extinction process). The difference between the two processes is clearly the presence of bias (towards zero population) in the extinction process, that was absent in the case of the random walk. This is a delicate and important point. We shall see that energy conservation, in thermally isolated systems, plays a role analogous to the requirement of zero bias, to guarantee entropy growth in the relaxation to equilibrium.

2.8 The Wiener-Khinchin Theorem

A rather natural way to characterize a random field, especially in the case of spatially homogeneous statistics, is to consider its Fourier spectrum. The same operation could obviously be carried out with a stochastic process, working in time.

Consider a random field $\psi(\mathbf{x})$, $\mathbf{x} \in \mathbf{R}^d$. We can define its Fourier transform:

$$\psi_{\mathbf{k}} = \int d^d x \, \psi(\mathbf{x}) e^{-i\mathbf{k}\cdot\mathbf{x}}; \qquad \psi(\mathbf{x}) = \int \frac{d^d k}{(2\pi)^d} \psi_{\mathbf{k}} e^{i\mathbf{k}\cdot\mathbf{x}}. \qquad (2.61)$$

From the mathematical point of view, it may be convenient to define the Fourier transform in terms of distributions; from the point of view of physics, a discretization and a finite domain for $\psi(\mathbf{x})$, will always be understood. Suppose that the statistics is spatially homogeneous, so that

$$\langle \psi(\mathbf{x})\psi(\mathbf{x}') \rangle = C(\mathbf{x} - \mathbf{x}')$$

In this case, the following relation will hold:

$$\langle \psi_{\mathbf{k}}\psi_{\mathbf{k}'} \rangle = \int d^d x \int d^d x' \, \langle \psi(\mathbf{x})\psi(\mathbf{x}') \rangle \exp[-i(\mathbf{k}\cdot\mathbf{x} + \mathbf{k}'\cdot\mathbf{x}')]$$
$$= \int d^d y \, C(\mathbf{y}) \exp(-i\mathbf{k}\cdot\mathbf{y}) \int d^d x' \, \exp[-i(\mathbf{k}+\mathbf{k}')\cdot\mathbf{x}'],$$

We notice that the integral $(2\pi)^{-d} \int d^d x' \exp[-i(\mathbf{k}+\mathbf{k}')\cdot\mathbf{x}']$ is just a representation of the Dirac delta $\delta(\mathbf{k}+\mathbf{k}')$. We find therefore

$$\langle \psi_{\mathbf{k}}\psi_{\mathbf{k}'} \rangle = (2\pi)^d \delta(\mathbf{k}+\mathbf{k}') C_{\mathbf{k}}, \qquad (2.62)$$

with

$$C_{\mathbf{k}} = \int d^d x \, C(\mathbf{x}) e^{-i\mathbf{k}\cdot\mathbf{x}}, \qquad (2.63)$$

the Fourier transform of the correlation function. This relation between correlations of Fourier components, $\psi_{\mathbf{k}}$, and the Fourier transform of the correlation, $C_{\mathbf{k}}$, is the content of the **Wiener-Khinchin theorem**. In several circumstances, it is possible to interpret $C(0) = \langle \psi^2 \rangle$, as an energy density. Then, the spectral decomposition provided by Eq. (2.62):

$$C(0) = \int \frac{d^d k}{(2\pi)^d} C_{\mathbf{k}}, \qquad (2.64)$$

allows to identify $C_{\mathbf{k}}$ as the **energy spectral density** of the random field (or of the signal).

The presence of the Dirac delta in Eq. (2.62), is the consequence of the infinite volume limit and of homogeneous statistics (homogeneous statistics implies that the random field does not decay to zero at infinity). In analogy to what was observed in the case of the random walker, additional singular behaviors may arise, due to non differentiability, or even lack of continuity of ψ. Such singular behaviors would imply divergence of quantities, such as $\langle |\nabla^p \psi|^2 \rangle$, with $p \geq 1$ in case of lack of continuity, $p \geq 2$ in case of non-differentiability, and so on. Writing in analogy with Eq. (2.64)

$$\langle |\nabla^p \psi|^2 \rangle \sim \int d^d k \, k^{2p} C_{\mathbf{k}}, \qquad (2.65)$$

we see that such divergent behaviors are associated with a power-law scaling at large k of the energy spectrum, $C_{\mathbf{k}} \sim k^{-d-2p+\zeta}$, with $\zeta \geq 0$.

From the point of view of physics, the divergence in Eq. (2.65) means simply that the value of $\langle |\nabla^p \psi|^2 \rangle$ is determined by the smallest scales a in the system, that provides an effective "ultraviolet" cutoff (at $k \sim a^{-1}$) in the integral.

We thus reach an alternative characterization of microscopic fluctuations, as fluctuations, whose spectrum decays slowly (or does not decay at all) at large k. An example is the instantaneous fine-grained density $\tilde{n}(\mathbf{x}, t) = \sum_i \delta(\mathbf{x} - \mathbf{x}_i(t))$. Its fluctuating part $\Delta \tilde{n}(\mathbf{x}, t) = \tilde{n}(\mathbf{x}, t) - \bar{n}$, has correlation $C(\mathbf{x} - \mathbf{y}) = \langle \Delta \tilde{n}(\mathbf{x}, t) \Delta \tilde{n}(\mathbf{u}, t) \rangle = \bar{n}\delta(\mathbf{x} - \mathbf{y})$ (see Sect. 3.3). Fourier transforming, we obtain trivially $C_{\mathbf{k}} = \bar{n}$, that in fact does not decay at large k.

We conclude illustrating how working with Fourier components, simplifies the coarse-graining operations illustrated in Sect. 2.6. The coarse-graining operation described in Eq. (2.51), was in fact in the form of a convolution. Hence, the Fourier transform of the coarse grained field, is just a filtered version of the original Fourier signal:

$$\psi_V(\mathbf{x}) = (G_V \psi)(\mathbf{x}); \qquad \psi_{V,\mathbf{k}} = \int d^d x \, \psi_V(\mathbf{x}) e^{-i\mathbf{k}\cdot\mathbf{x}} = G_{V,\mathbf{k}} \psi_{\mathbf{k}}, \qquad (2.66)$$

where the condition $\int d^d x \, G_V(\mathbf{x}) = 1$ gives us $G_{V,\mathbf{k}\to 0} = 1$, while, for $kV^{1/d} \gg 1$, we must have $G_{V,\mathbf{k}} \to 0$.

2.9 Problems

Problems 1 The two random variable x and y are distributed in the square $[0, 1] \otimes [0, 1]$ with the law $\rho(x, y) = C(1 - x - y)\theta(1 - x - y)$, with C a constant and θ the Heaviside step function $(\theta(z > 0) = 1, \theta(z < 0) = 0)$. Calculate the conditional PDF $\rho(x|y > 0.5)$ and the conditional probability $P(y > 0.5|x)$.

Problems 2 A rather poor target shooter, the few times that hits the target, hits it at a random position. Suppose the target to be a radius R circle. Calculate the PDF that a hit on target, occurs at radius r. Calculate the probability that a hit on target, falls at $r < R/2$.

Problems 3 The velocity PDF for a molecule of a monoatomic ideal gas at thermodynamic equilibrium, is given by the Maxwell distribution

$$\rho_1(\mathbf{v}) = \left(\frac{m}{2\pi KT}\right)^{3/2} \exp\left(-\frac{mv^2}{2KT}\right), \tag{2.67}$$

where m is the mass of the molecule, T is the temperature and K is the Boltzmann constant. Indicate by N the number of molecules in the gas, and by V the volume in which it is contained.

- Write down the one-molecule PDF $\rho_1(\mathbf{x}, \mathbf{v})$, and the joint PDF, for all the molecules in the gas $\rho_N(\{\mathbf{r}_i, \mathbf{v}_i\})$.
- Calculate the PDF entropy $S = S(T, V, N) = -\langle \ln \rho_N \rangle$. How does it compare with standard expressions for the thermodynamic entropy of a gas?
- Estimate numerically the internal energy $E = \sum_{i=1}^{N} \frac{1}{2}m|\mathbf{v}_i|^2$, and the amplitude of its fluctuations, given values of the parameters $K \simeq 1.4 \cdot 10^{-23} \mathrm{J\,K^{-1}}$, $N \sim 10^{20}$, $V \simeq 1\mathrm{m}^3$, $T \sim 300\,\mathrm{K}$, $m \simeq 7 \cdot 10^{-27}\,\mathrm{Kg}$.

Problems 4 Adapt the derivation of the evolution equations for the moments x^n, described in Sect. 2.7.1, to calculate the evolution of the PDF entropy of the walker. Prove that Eq. (2.56) leads to an entropy that is a monotonously increasing function of time. Repeat the calculation for the extinction process, and verify that entropy growth, in that case, is not guaranteed. Provide an example of a situation of (temporary) entropy growth for an extinction process.

2.10 Further Reading

An elementary presentation of probability theory, with lots of useful exercises is contained in:

- S. Lipschutz, *Theory and Problems of Probability* (Schaum outline series, McGraw Hill, 1968. Freely available online)

The reference book for probability and basic stochastic process theory, however, is:

- W.Feller, *An Introduction to Probability Theory and Its Applications* (Wiley, 1968)

More on the concept of entropy and on information theory:

- A. Papoulis, S.U. Pillai, *Probability, Random Variables and Stochastic Processes* (McGraw Hill, 2002)

Good reference books for stochastic processes:

- C. Gardiner, *Handbook of Stochastic Methods* (Springer, 1986)
- P.E. Kloeden, E. Platen, *Numerical Solution of Stochastic Differential Equations* (Springer, 1992)
- R.L. Stratonovich, *Topics in the Theory of Random Noise. Volume I* (Gordon and Breach, 1963)

I have taken the material for anomalous diffusion from:

- J.-P. Bouchaud, A. Georges, Anomalous diffusion in disordered media: statistical mechanisms, models and physical applications. Phys. Rep. **195**, 127 (1990)

More on this also in:

- D. Sornette, *Critical Phenomena in Natural Sciences* (Springer, 2000)

Appendix

A.1 The Central Limit Theorem

We want to determine the form of the probability distribution for sums of independent identically distributed random variables (i.i.d.) in the form

$$X_N = \sum_{k=1}^{N} x_k.$$

A fundamental question is the existence of limit forms at large N for the PDF $\rho(X_N)$. Such limit distributions indeed exist, and their form depends solely on the behavior of the PDF $\rho(x)$ at large values of the argument: the so-called **tails of the distribution**. In particular, the existence of the first moments of the distribution is crucial in the determination of the limit form for $\rho(X_N)$.

We recall that the $n$the moment of a distribution will exist,

$$\langle x^n \rangle = \int dx\, x^n \rho(x) < \infty,$$

provided $\rho(x)$ goes to zero at infinity faster than x^{-1-n}.

We can distinguish, substantially, three cases:

- Both the variance σ_x^2 and the mean μ_x are finite, that is the situation, assumed in Sect. 2.3, for a thermodynamic limit to occur. In this case, $\rho(X_N)$ will be a gaussian distribution, with mean $\mu_{X_N} = N\mu_x$ and variance $\sigma_{X_N}^2 = N\sigma_x^2$.
- The mean μ_x is finite, but the variance σ_x^2 is infinite, meaning that, for $x \to \infty$, $\rho(x) \sim x^{-1-\alpha}$ (within logarithms), with $1 < \alpha < 2$. In this case, again, $\mu_{X_N} = N\mu_x$, but the deviations $\hat{X}_N = X_N - \mu_{X_N}$ are distributed with a so-called Lévy law of index α, whose asymptotic behavior at large $\hat{X}_N$ is $\rho(\hat{X}_N) \sim \hat{X}_N^{-1-\alpha}$.
- Also the mean μ_x is infinite. This is typically realized by an asymmetric distribution $\rho(x)$, with asymptotic behavior at large x: $\rho(x) \sim x^{-1-\alpha}$, with $0 < \alpha < 1$ (for $\alpha < 0$, the PDF would not be normalized). In this case, the limit distribution for X_N would be a Lévy law of index α, whose asymptotic behavior at large X_N is $\rho(X_N) \sim X_N^{-1-\alpha}$.

The form of the limit distribution, can be calculated, exploiting the important property that the characteristic function of a sum of random variables, is the product of the characteristic functions of the addends in the sum. The PDF of a sum of independent random variables, is in fact the convolution of the PDF's of the addends:

$$\rho_{x+y}(z) = \int dy \, \rho_x(z-y)\rho_y(y),$$

and the characteristic function of $x+y$, being the Fourier transform of a convolution, will be the product

$$Z_{x+y}(j) = Z_x(j)Z_y(j).$$

Thus, in the case of a sum of i.i.d. random variables:

$$Z_{X_N}(j) = (Z_x(j))^N. \tag{2.68}$$

This is the quantity on which we shall focus, to determine the asymptotic behavior, for $N \to \infty$, of the PDF $\rho(X_N)$.

A.1.1 The Gaussian Case

If the mean and the variance of the random variable x are both finite, we know from the law of large numbers that $\mu_{X_N} = N\mu_x$ and $\sigma_{X_N}^2 = N\sigma_x^2$. We scale out the dependence of the PDF $\rho(X_N)$ on the parameters μ_{X_N} and σ_x^2, by considering a rescaled version of the deviation from the mean $\hat{X}_N = X_N - \mu_{X_N}$: $Y_N = \hat{X}_N/\sigma_{X_N}$. We verify that the limit $\rho = \lim_{N\to\infty} \rho_{Y_N}$ exists, and is indeed the Gaussian.

We have for the characteristic function of the rescaled variable Y_N:

$$Z_{Y_N}(j) = \int dY_N \rho_{Y_N}(Y_N)(ijY_N)$$

$$= \int d\hat{X}_N \, \rho_{\hat{X}_N}(\hat{X}_N) \exp(ij\hat{X}_N/\sigma_{X_N}) = Z_{\hat{X}_N}(j/\sigma_{X_N}), \qquad (2.69)$$

and, from Eq. (2.68):

$$Z_{Y_N}(j) = (Z_{\hat{x}}(j/\sigma_{X_N}))^N, \qquad (2.70)$$

where $\hat{x} = x - \mu_x$. Taking the $N \to \infty$ limit in ρ_{Y_N}, corresponds to taking the limit $j/\sigma_{X_N} \to 0$ in $Z_{\hat{x}}(j/\sigma_{X_N})$. We can proceed by Taylor expansion. Since $\langle \hat{x} \rangle = 0$, the Taylor expansion of $Z_{\hat{x}}$, does not contain the linear term:

$$Z_{\hat{x}}(j) = 1 - \frac{1}{2}\sigma_x^2 j^2 + o(j^2). \qquad (2.71)$$

Hence, substituting into Eq. (2.70), and using $\sigma_{X_N}^2 = N\sigma_x^2$:

$$Z(J) = \lim_{N\to\infty} Z_{Y_N}(j) = \lim_{N\to\infty} \left(1 - \frac{j^2}{2N} + o(N^{-1})\right)^N = \exp(-j^2/2). \qquad (2.72)$$

Inverse Fourier transforming, we find that the limit of the distribution for Y_N is the Gaussian

$$\rho(Y_N) = \frac{1}{(2\pi)^{1/2}} \exp(-Y_N^2/2), \qquad (2.73)$$

as claimed.

The relevance of this result to the random walk dynamics, described in Sect. 2.7.1, should be apparent. If we attach a time label $t_k = k\Delta t$ to each random variable x_k, and take $\mu_x = 0$, the sum X_N will be precisely the displacement of a random walker, that, in the time $t_N = N\Delta t$, has performed N independent steps x_k. What Eq. (2.73) tells us, is that the density profile of a population of walkers, starting from a common initial position, will have, after a sufficiently long time (provided the domain is infinite), a Gaussian profile. This is in fact the result in Eq. (2.57), where the $N \to \infty$ limit was obtained implicitly with the continuous limit $t/\Delta t \to \infty$.

We have proved that, for fixed Y, the PDF $\rho_{Y_N}(Y)$ has the limit $\rho(Y)$, given in Eq. (2.73). The natural question then arises about the range of Y_N for which the result in Eq. (2.73), for large *but finite* N, holds. For sure, $\rho_{Y_N}(Y)$ will begin to be sensitive to the properties of the tails of ρ_x, when $X_N - \mu_{X_N} \sim N\sigma_x$, i.e. when $Y_N \sim N^{1/2}$. For instance, if $\rho(x) = 0$ for $|x - \mu_x| > \Delta x$ (as in the case of the jump distribution of a random walker), ρ_{Y_N} will surely be zero for $|Y_N| > N^{1/2}$. In other words, the central limit result of Gaussian statistics, for a large but finite sum of i.i.d.

random variables, with finite mean and variance, will hold only far from the tails of the distribution.

We can provide a quantitative estimate of this effect, in the case the first correction to $Z_{\hat{x}}$ is a non-zero third moment $\langle \hat{x}^3 \rangle \neq 0$:

$$Z_{\hat{x}}(j) = 1 - \frac{1}{2}\sigma_x^2 j^2 - \frac{i}{6}\langle \hat{x}^3 \rangle j^3 + o(j^3).$$

Substituting into Eq. (2.70), we are able to include the leading large N correction in Eq. (2.72):

$$Z_{Y_N}(j) = \left(1 + i\alpha s_3 N^{-1/2} j^3 + o(N^{-1/2})\right) \exp(-j^2/2),$$

where α is a numerical coefficient, and $s_3 = \langle \hat{x}^3 \rangle / \sigma_x^{3/2}$ is the normalized third moment (so-called skewness) of the distribution $\rho(\hat{x})$. Inverse Fourier transforming, we find the correction to Eq. (2.73):

$$\frac{\rho_{Y_N} - \rho}{\rho} \sim s_3 N^{-1/2} Y_N^3. \tag{2.74}$$

In order for the central limit to hold, we need that $|Y_N| \ll s_3^{1/3} N^{1/6}$, i.e. $|X_N - \mu_{X_N}| \ll s_3^{1/3} N^{2/3}$.

A.1.2 Lévy Distributions

We pass to consider the case in which the PDF $\rho(x)$ does not have its first or second moment. We have seen that this corresponds to a power law behavior in the tails of the distribution: $\rho(x) \sim x^{-1-\alpha}$, with $0 < \alpha < 1$ in the first case, $1 < \alpha < 2$ in the second. The presence of a power law in the tails of the distribution, $\rho(x) \sim x^{-1-\alpha}$, will be associated, in general, with non-existence of moments $\langle x^n \rangle$ with $n > \alpha$. Thus, the Taylor expansion in $j = 0$ of the characteristic function Z_x, will stop at $\mathrm{int}(\alpha)$, and it is possible to see that the remnant in the Taylor expansion is in the form $c|j|^\alpha$, with c a numerical constant. If $\alpha > 2$, this non-analyticity will not modify the limiting form of $Z_{\hat{x}}$, given in Eq. (2.71). If, on the other hand, either $0 < \alpha < 1$, or $1 < \alpha < 2$ and $\mu_x = 0$, this form must be replaced by

$$Z_x(j) \simeq 1 - c|j|^\alpha. \tag{2.75}$$

We thus see that, contrary to the finite μ_x and σ_x^2 case, the behavior of the tails of the PDF ρ_x is reflected in the behavior, for $j \to 0$, of the characteristic function Z_x.

To determine the PDF ρ_{X_N}, we resort again to the method of characteristic functions. We find immediately the result $Z_{X_N}(j) = (1 - c|j|^\alpha)^N$, which suggests us that, in order to obtain a limit distribution, it is necessary to rescale X_N. From Eq. (2.69), we see that the proper rescaling is

$$Y_N = N^{-\alpha} X_N.$$

This leads to the asymptotic form of the characteristic function

$$Z(J) = \lim_{N \to \infty} Z_{Y_N}(j) = \lim_{N \to \infty} \left(1 - \frac{c|j|^\alpha}{N} + o(N^{-1})\right)^N = \exp(-c|J|^\alpha). \quad (2.76)$$

The limit distribution $\rho(Y_N)$, obtained inverse Fourier transforming Eq. (2.76), is called a **Lévy distribution** (or **stable distribution**) of order α. Its most important property, revealed by the non-analyticity of Z in $j = 0$, is the power law behavior in the tails of $\rho(Y_N)$: $\rho(Y_N) \sim Y_N^{-1-\alpha}$, that is the same as for the individual variable x. The fact that Y_N, that is a sum of infinite-mean i.i.d. random variables, is still an infinite-mean random variable, is not surprising. The fact that the scaling behavior, in the tails of ρ_x and $\rho = \lim_{N \to \infty} \rho_{Y_N}$, is identical, reflects the different origin of the limit behavior of the ρ_{Y_N} in the Gaussian and in the Lévy case. While in the Gaussian case, the form of the limit distribution does not probe the tails of ρ_x, in any way other than existence of the first moments of the distribution, in the Lévy case, it is precisely the scaling in the tails that determines the form of the limit distribution. In fact, we can interpret the law of large number result, $\langle x \rangle_N \to \mu_x$, as a manifestation of the fact that each x_k contributes to X_N a term of the same order $\sim \mu_x$. This property must apparently be lost in the $\mu_x \to \infty$ case. What happens is that the value of X_N is determined typically by the largest x_k in the sequence $\{x_k, k = 1, \ldots, N\}$.

We can get a quantitative feeling of this phenomenon, from the observation that the values of x_k, in a typical sequence $\{x_1, x_2, \ldots, x_N\}$, will be distributed with the PDF ρ_x. We see that, in order for a certain large value $\bar{x}$ (or larger), in a typical sequence of N, to be observed, it is necessary that $NP(x > \bar{x}) \gtrsim 1$, i.e.:

$$\int_{\bar{x}}^{\infty} dx \, \rho_x(x) \sim x^{-\alpha} \gtrsim 1/N.$$

Thus, the largest value of x observed in a typical sequence $\{x_1, x_2, \ldots, x_N\}$ will be:

$$x_N^{MAX} \sim N^{1/\alpha}.$$

All the smaller x_k's in the typical sequence, will be distributed with ρ_x. Hence, we can estimate

$$X_N \sim N \int_0^{x_N^{MAX}} dx \, x \rho_x(x) \sim N(X^{1/\alpha})^{1-\alpha} = N^{1/\alpha}.$$

The sum X_N scales with the largest typical contribution x_N^{MAX}, meaning that it is the largest contribution in the sequence $\{x_k, k = 1, \ldots, N\}$, that dominates X_N.

As we have done in the Gaussian case, we can map the problem of summing i.i.d. random variables, to a random walk, by attaching a time label $t_k = k\Delta t$ to each increment x_k. To have an unbiased random walk, we need $\mu_x = 0$, so that the interesting dynamics is the one originating from the $1 < \alpha < 2$ regime of infinite variance—zero mean increments. The result is an infinite variance displacement X_N, in time $t_N = N\Delta t$. The resulting process goes under the name of **Lévy flight**, and could be used to describe migration of individuals, that have at their disposal means of locomotion of very diverse nature (think of human beings that, in a single day, can move by few meters, as well as embark on an intercontinental flight). The resulting evolution equation in the continuous limit, is an example of the non-local master equation (2.54), in which the propagation kernel is precisely the single increment distribution: $w(x \to y) \propto \rho(x - y) \sim |x - y|^{-1-\alpha}$.

Additional discussion of these issues can be found e.g. in [J.-P. Bouchaud and A. Georges, Anomalous diffusion in disordered media: statistical mechanisms, models and physical applications, Phys. Rep. **195**, 127 (1990)] and in [W. Feller, An introduction to probability theory and its applications (Wiley and Sons, 1968), Vol. 1].

Chapter 3
Kinetic Theory

3.1 From Γ-Space to Macroscopic Distributions

The general idea behind kinetic theory and statistical mechanics is to obtain the macroscopic properties of a physical system from a probabilistic description of its microscopic dynamics. Broadly speaking, we can say that statistical mechanics deals with systems in thermodynamic equilibrium, while kinetic theory is more concerned with time-dependent non-equilibrium conditions.

We have seen in Sect. 2.3 that the observed values of macroscopic variables are actually averages of fluctuating quantities. At thermodynamic equilibrium, these fluctuations are a consequence of the discrete nature at microscopic scales of the system, and become negligible in the thermodynamic limit (except, as we shall discuss in Sect. 4.11.2 at critical points). Far from thermal equilibrium, fluctuations in general will not be negligible (think of turbulence), so that the coincidence between observed and average values of the macroscopic variables, will simply disappear. The general idea of a connection between macroscopic properties of a physical system, and statistical properties of its microscopic parts, however, remains valid. Even more strongly, we shall see that simple averages (albeit with an interpretation different from the one at equilibrium) remain central in the description of the systems.

We shall denote the state of a system formed by N microscopic parts by the letter Γ, from which the name of Γ-**space** for the microscopic phase space of the system. In the case of a simple monoatomic gas, $\Gamma = (\mathbf{y}_1, \mathbf{y}_2, \ldots, \mathbf{y}_N)$, with $\mathbf{y}_k = (\mathbf{x}_k, \mathbf{v}_k)$ the vector formed from the position $\mathbf{x}_k$ and velocity $\mathbf{v}_k$ of the k-th molecule. We can introduce the PDF $\rho_N(\Gamma, t)$ of the microscopic degrees of freedom of the system and use it to calculate averages of macroscopic variables $\tilde{m}(\Gamma(t))$. In thermal equilibrium, we expect that the most probable value of the variable, $m(t)$, and its average with respect to $\rho_N(\Gamma, t)$, $\langle \tilde{m}(\Gamma(t)) \rangle$, will coincide.

A typical situation of interest in kinetic theory is the relaxation to equilibrium of the macroscopic variable $m(t)$, from an initial non-equilibrium condition $m(0) = m_0$. To fix the ideas, consider again the system of Sects. 2.3 and 2.4: a gas of N molecules in a volume V, where we take, as macroscopic variable, the number of

© Springer International Publishing Switzerland 2015

P. Olla, *An Introduction to Thermodynamics and Statistical Physics*,
UNITEXT for Physics, DOI 10.1007/978-3-319-06188-7_3

molecules in a subvolume $V_a \in V$. The instantaneous value of the variable will be $\tilde{m}(\Gamma(t)) \equiv \tilde{N}_a(\Gamma(t)) = \sum_{k=1}^{N} \delta_{V_a}(\mathbf{x}_k)$, where δ_{V_a} is the indicator function of the volume V_a; the corresponding observed value will be the average:

$$m(t) = \int d\Gamma \, \tilde{m}(\Gamma) \rho_N(\Gamma, t | m(0) = m_0) \equiv \langle \tilde{m}(\Gamma(t)) | m(0) = m_0 \rangle. \tag{3.1}$$

We can understand Eq. (3.1) as the limit of a sample average, in a sequence of experiments in which the system is prepared initially in the same macroscopic condition $m(0) = m_0$. In the system of Sect. 2.3, $m(0) = m_0$ could represent a local unbalance in the density of the gas, in a situation in which all the other macroscopic variables in the system (temperature, density in other points of the tank) are supposed at equilibrium. Such a collection of identicaly systems, each in a distinct microscopic state, but sharing the same values of the macroscopic variables is typically referred to with the name of **ensemble**.

If we let the system evolve, we will see that in each experiment a different outcome $\tilde{m}(\Gamma(t))$ is produced. The reason is that the condition $m(0) = m_0$ does not fix the initial microscopic state $\Gamma(0)$. Thus, different initial conditions will lead to different evolutions $\Gamma(t)$ and different values of $\tilde{m}(\Gamma(t))$ (depending on the system, the evolution of Γ may be even stochastic itself). If macroscopic fluctuations are absent, however, all the $\tilde{m}(\Gamma(t))$ will group close to $m(t)$ in the thermodynamic limit. (We shall return to the concept of macroscopic fluctuations later in the chapter).

A distribution such as $\rho_N(\Gamma, t)$, whose domain of definition has dimensionality of the order of the Avogadro number, is clearly a very difficult object to deal with. Fortunately, most macroscopic variables $\tilde{m}$ can be expressed as sums of microscopic variables. This means that, to determine $m(t)$, it is sufficient to know the marginal one-molecule distributions $\rho_{1,k}(\mathbf{y}_k, t)$. A further simplification will arise if the molecules are identical, or, more in general, if the system dynamics is invariant under exchange between microscopic components. A PDF $\rho_N(\Gamma, t)$ that is initially symmetric under exchange of any two variables $\mathbf{y}_k$ and $\mathbf{y}_j$, will in this case remain symmetric at all later times (invariant dynamics means that the evolution of the system is the same in the two cases). All the marginals $\rho_1(\mathbf{y}; t), \rho_2(\mathbf{y}, \mathbf{y}'; t), \ldots$ will have therefore identical form.

Back to the system of Sects. 2.3 and 2.4, this leads to the result:

$$N_a(t) = N P(\mathbf{x}_k \in V_a, t). \tag{3.2}$$

We can generalize the expression to macroscopic quantities that are sums of products of microscopic variables: an example is the interaction energy between a pair of molecule $U(\mathbf{x}_k - \mathbf{x}_j)$; the corresponding contribution to the internal energy of the gas would be $\frac{N(N-1)}{2} \langle U(\mathbf{x}_k - \mathbf{x}_j) \rangle$, that is calculated from a two-molecule PDF $\rho_2(\mathbf{y}_k, \mathbf{y}_j; t)$, that has the same form $\forall k \neq j$.

From Eq. (3.2) we can define a number of quantities of fundamental importance for the later development of the theory. In first place the mean density $n_{V_a}(\mathbf{x}_a, t) = N_a(t)/V_a$, where $\mathbf{x}_a$ is the point around which the volume V_a is centered. If V_a is small on the scale of variation of $\rho_1(\mathbf{x}, t)$, we can write in Eq. (3.2) $P(\mathbf{x}_k \in V_a, t) \simeq \rho_1(\mathbf{x}_k, t)V_a$, and the mean density becomes independent of V_a:

$$n(\mathbf{x}, t) = N\rho_1(\mathbf{x}, t). \tag{3.3}$$

We can generalize Eq. (3.3) to the full phase space for one molecule, and define the so called **distribution function** for the gas:

$$f_1(\mathbf{x}, \mathbf{v}; t) = \frac{dN(\mathbf{x}, \mathbf{v}; t)}{d^3x d^3v} = N\rho_1(\mathbf{x}, \mathbf{v}; t). \tag{3.4}$$

The distribution function f is the central object of study of kinetic theory, out of which, any macroscopic quantity, expressible as a sum of one-molecule contributions, could be determined. We point out here, that we have passed from a microscopic description, such as the one provided by $\rho_N(\Gamma, t)$ on the state of each individual molecule in the system, to a macroscopic one, in which the quantity of interest is a number of molecules.

We can introduce the instantaneous counterparts to the mean quantities $n(\mathbf{x}, t)$ and $f_1(\mathbf{y}; t)$, working with indicator functions:

$$\tilde{f}_{1,\Delta y}(\mathbf{y}; t) = \frac{1}{\Delta y} \sum_{k=1}^{N} \delta_{\Delta y}(\mathbf{y}_k), \tag{3.5}$$

where, again, $\mathbf{y}$ is the point in the phase space of an individual microscopic part, around which the volume Δy is centered. Analogous formulae can be derived for $\tilde{n}_{\Delta V}(\mathbf{x}, t)$. We notice that the $\Delta y \to 0$ limit of Eq. (3.5) is singular, as the ratio $\delta_{\Delta y}(\mathbf{y}_k)/\Delta y$ tends to the Dirac delta $\delta(\mathbf{y}_k - \mathbf{y})$. The $\Delta y \to 0$ limit of Eq. (3.5) would be in fact

$$\tilde{f}_1(\mathbf{y}; t) = \sum_{k=1}^{N} \delta(\mathbf{y}_k(t) - \mathbf{y}), \tag{3.6}$$

that is sometimes called the **Klimontovich distribution**. It is possible to see that the Klimontovich distribution provides the same amount of information contained in a Γ-space PDF in the form $\rho_N(\bar{\Gamma}, t) = N!^{-1} \sum_{perm.} \delta(\Gamma(t) - \bar{\Gamma})$, where the permutations in the sum, are carried out on the labels in the vectors appearing in $\bar{\Gamma} = (\mathbf{y}_1, \mathbf{y}_2, \dots)$. This is the same amount of information to determine the position and velocities of all the molecules in a gas, without specifying which molecule has the given position and velocity.

3.2 Statistical Closure

In order for a macroscopic system to have a meaningful dynamics, it is necessary that its microscopic constituents interact in some way. This leads to statistical dependence among the microscopic degrees of freedom. Thus, it is to be expected that it is not possible to derive, in closed form, evolution equations for f_1. We shall see that resolution of the evolution equation for any distribution ρ_k, with $k < N$, would require knowledge of PDF's ρ_j, taking into account a number $j > k$ of parts of the system. This is the so called **closure problem**: because of interactions, the evolution of the k molecules described by ρ_k does not depend solely on the k molecules themselves, but also on all the others that are present in the system.

We illustrate the situation with a simple toy model: a set of N interacting random walkers moving on the line. Let Δx be their (discrete) hopping length, and Δt the associated discretization in time. The walkers interact by modifying the hopping rates $w_J \equiv P_J/\Delta t$ of the walker near to them. The following expression is assumed:

$$\tilde{w}_J(x_i, t) = \sum_{j \neq i}^{N} \hat{q}(x_i - x_j(t)), \tag{3.7}$$

where $\hat{q}(x)$ is a positive defined symmetric function, with support in the interval $[-\lambda, \lambda]$, with $\Delta x \ll \lambda$. Thus, a walker will have the possibility to hop only if there is another walker at distance $< \lambda$: λ plays the role of an effective interaction length for the system.

To determine the evolution of $\rho_1(x_i, t)$, and therefore of $f_1(x, t)$, we must average Eq. (3.7) over the position of all the other walkers $j \neq i$:

$$w_J(x, t) = \left\langle \sum_{j \neq i}^{N} \hat{q}(x_i(t) - x_j(t)) \Big| x_i(t) = x \right\rangle,$$

which can be rewritten in the form

$$w_J = \sum_{j \neq i} \int_0^L dz \hat{q}(x - z) \langle \delta(x_j(t) - z) | x_i(t) = x \rangle = (N - 1) \int_0^L dz \hat{q}(x - z) \rho_2(z, t | x, t).$$

To determine the evolution equation for ρ_1, we follow the same procedure that we have utilized to go from Eq. (2.55) to (2.56). The result is:

$$\frac{\partial \rho_1(x, t)}{\partial t} = \frac{\partial(w_J(x, t)\rho_1(x, t))}{\partial x} = (N - 1)\frac{\partial^2}{\partial x^2} \int_0^L dz \, q(x - z)\rho_2(z, t | x, t)\rho_1(x, t),$$

where $q = \hat{q}(\Delta x)^2$. Multiplying by N, we find:

$$\frac{\partial f_1(x, t)}{\partial t} = \frac{\partial^2}{\partial x^2} \int_0^L dz \, q(x - z) f_2(z, x; t). \tag{3.8}$$

In order to determine f_1, we must know f_2, as expected. If we tried to write the equation for f_2, we would quickly discover that it requires knowledge of f_3. And so on, the equation for f_k requires knowledge of f_{k+1}, till the equation for f_N, which, by construction, is closed in itself and does not require knowledge of any higher order distribution. The resulting hierarchy of equations, typically referred to with the name **BBGKY hierarchy**, is thus not closed at any level $k < N$.

Short of solving the whole hierarchy, or, equivalently, trying to solve the equation for f_N, the only possibility is to resort to some kind of approximation. A possibility, called **statistical closure** approach, is to impose the functional dependence between the f_j's, with j above some fixed k, on the f_j's with $j \le k$. This guarantees that we can cut the BBGKY hierarchy at any order $j \ge k$, and the resulting system of equations will be closed.

An example of closure is to assume Gaussian statistics, which corresponds to the choice $k = 2$. The most severe choice, of course, is $k = 1$, that is equivalent to assuming that f_N fully factorizes in products of f_1's (equivalently, that ρ_N factorizes in products of ρ_1's). In particular:

$$f_2(x, z; t) \simeq f_1(x, t) f_1(z, t). \tag{3.9}$$

This assumption goes under the name of **mean field approximation**, and corresponds in fact to disregarding the fluctuations $\tilde{f}_1 - f_1$ in the dynamics.

Utilizing the mean field approximation Eq. (3.9) in Eq. (3.8), we obtain the non-linear diffusion equation

$$\frac{\partial f_1(x, t)}{\partial t} = \frac{\partial^2}{\partial x^2} \int dz \, q(x - z) f_1(x, t) f_1(z, t),$$

and, if f_1 varies at scale $\gg \lambda$:

$$\frac{\partial f_1(x, t)}{\partial t} = \alpha \frac{\partial^2 f_1^2(x, t)}{\partial x^2}, \qquad \alpha = \int_{-\infty}^{+\infty} q(x) dx. \tag{3.10}$$

Summarizing:

- Interactions lead to an unclosed hierarchy of linear equations for the distributions f_k, $k = 1, 2, \ldots, N$. Specifically, a two-body interaction, such as the one in Eq. (3.7), will lead to equations for f_k, requiring knowledge of the distribution f_{k+1}. It is possible to see that 3-body interactions would require knowledge of f_{k+2} and so forth.

- Statistical closure at a certain level k in the hierarchy, will lead to a system of k nonlinear equations.

Note It is possible to show that the dynamics described in Eq. (3.10) leads to monotonic growth of the entropy associated with the PDF $\rho_1(x, t) = f_1(x, t)/N$: $S_1 \equiv S[\rho_1 \delta x] = -\int_0^L dx\, \rho_1(x, t) \ln(\rho_1(x, t)\delta x)$. Let us consider, for simplicity, the case in which the domain is a ring (periodic boundary conditions) of length L. We have, from Eq. (3.10):

$$\dot{S}_1 = -\frac{d}{dt} \int_0^L dx \rho_1(x, t) \ln(\rho_1(x, t)\delta x) = -\int_0^L dx \dot{\rho}_1(x, t) \ln(\rho_1(x, t)\delta x). \quad (3.11)$$

The kinetic equation for ρ_1 coincides in form with Eq. (3.10); substituting into Eq. (3.11):

$$\dot{S}_1 = -\alpha \int_0^L dx \frac{\partial^2 \rho_1^2(x, t)}{\partial x^2} \ln(\rho_1(x, t)\delta x).$$

Integrating by part, the boundary terms cancel and we remain with:

$$\dot{S}_1 = \alpha \int_0^L dx \frac{1}{\rho(x, t)} \frac{\partial \rho_1^2(x, t)}{\partial x} \frac{\partial \rho_1(x, t)}{\partial x} = 2\alpha \int_0^L dx \left(\frac{\partial \rho_1(x, t)}{\partial x}\right)^2 \geq 0.$$

Entropy grows, as expected.∎

3.3 The Role of Fluctuations

We have already hinted to the fact that there exist basically two kinds of fluctuations: microscopic fluctuations, that disappear in the thermodynamic limit, and macroscopic ones, that do not. Microscopic fluctuations arise from the discrete nature of the system (the existence of discrete microscopic components). Macroscopic fluctuations may be present in some special thermodynamic equilibrium conditions (critical systems) and in non-equilibrium conditions (turbulence). They arise due to correlations in the system, that extend up to macroscopic scales.

We can convert the statement on correlations between microscopic component, into one on correlations between macroscopic quantities, such as the particle density, or some other quantity with identical meaning. Due to the presence of fluctuations, these quantities will have the structure of random fields. We shall see that, in order for fluctuation to be considered microscopic, it is necessary that their correlation length be zero in the continuum limit.

As an example of microscopic fluctuations, let us consider, as macroscopic variable, the molecule count $\tilde{N}_a$ in a volume V_a, already considered in the previous section. We must be careful in taking the limit $V_a \to 0$, due to the presence of Dirac deltas in expressions such as Eq. (3.6). Let us then calculate the density fluctuation correlation first, and take the $V_a \to 0$ limit afterwards. If $\tilde{N}_{a,b} = \sum_k \delta_{V_{a,b}}(\mathbf{x}_k(t))$ are the instantaneous particle counts in the volumes $V_{a,b}$ centered around the points $\bar{\mathbf{x}}_{a,b}$, the coarse-grained densities $\tilde{n}_{V_{a,b}}(\bar{\mathbf{x}}_{a,b}, t)$ will have correlation

$$\langle \tilde{n}_{V_a}(\bar{\mathbf{x}}_a, t) \tilde{n}_{V_b}(\mathbf{x}_b, t) \rangle = n(\mathbf{x}_a, t)n(\mathbf{x}_b, t) + (\delta_{ab}/V_a)n(\mathbf{x}, t). \qquad (3.12)$$

To understand the origin of the additional term for $a = b$, recall the expression $\langle \tilde{N}_a^2 \rangle = \langle (\sum_k \delta_{V_a}(\mathbf{x}_k))^2 \rangle = N(N-1)\langle \delta_{V_a} \rangle^2 + N\langle \delta_{V_a} \rangle$ (see Sect. 2.3). Taking the $V_{a,b} \to 0$ limit, we obtain the result

$$\langle \tilde{n}(\bar{\mathbf{x}}_a, t) \tilde{n}(\mathbf{x}_b, t) \rangle = n(\mathbf{x}_a, t)n(\mathbf{x}_b, t) + n(\mathbf{x}_a, t)\delta(\mathbf{x}_a - \mathbf{x}_b). \qquad (3.13)$$

The effective width of the Dirac delta in Eq. (3.13), is given by the discrete scale $n^{-1/3}$, that corresponds to the typical molecular separation. This allows to interpret the thermodynamic limit condition in volumes of size ΔV, $n\Delta V \gg 1$, as an ergodic properties for the volume averages of the density $\tilde{n}$:

$$\tilde{n}_{\Delta V}(\mathbf{x}, t) = \langle \tilde{n}(\mathbf{x}, t) \rangle_{\Delta V} \simeq \langle \tilde{n}(\mathbf{x}, t) \rangle \equiv n(\mathbf{x}, t). \qquad (3.14)$$

The mean density $n(\mathbf{x}, t)$ could then be seen as the volume average of the instantaneous density $\tilde{n}$, over volumes ΔV large enough for the thermodynamic limit to be satisfied, but small compared with the scale L of variation for n (a condition for n to be considered a macroscopic quantity, obviously, is that its scale of variation is macroscopic: $L \gg n^{-1/3}$).

The correlation length for the fluctuations is a crucial parameter to decide whether a mean-field approximation can be utilized or not. If one were to judge simply on the basis of the fluctuation amplitude, the Dirac deltas present in Eq. (3.6), and the fact that $\tilde{n} - n$ is itself a sum of Dirac deltas, would suggest that a mean-field approximation would never be applicable. But this is not the case. We can verify that density fluctuations with the correlation structure described by Eq. (3.13) are harmless, on the example of the interacting random walkers described in the previous section. Taking in Eq. (3.8), $f_2(z, x; t) = f_1(z, t)f_1(x, t) + f_1(x, t)\delta(x - z)$, would give the deviation from mean field theory:

$$\int_0^L dz \, q(x-z)[f_2(z, x; t) - f_1(z, t)f_1(x, t)] = q(0)f_1(x, t), \qquad (3.15)$$

while

$$\int_0^L dz \, q(x-z)f_1(z, t)f_1(x, t) \sim (\lambda f_1(x, t))q(0)f_1(x, t). \qquad (3.16)$$

But λf_1 is the number of walkers in the interaction length λ. Thus, the condition for a mean-field theory to be applicable, is that the thermodynamic limit be applicable at scale λ, i.e. $\lambda f_1 \gg 1$.

We can examine the opposite limit of macroscopic fluctuations with correlation length $\lambda_c \gg \lambda$. This means that the fluctuation correlation $f_{2c}(z, x; t) = f_2(z, x; t) - f_1(z, t)f_1(x, t)$ varies little at scale λ, so that we have, in place of Eq. (3.15):

$$\int_0^L dz \, q(x - z)[f_2(z, x; t) - f_1(z, t)f_1(x, t)] \sim q(0)\lambda f_{2c}(x, x; t).$$

Comparing with Eq. (3.16), we see that the condition for the mean-field approximation to be applicable, becomes $f_{2c} \ll f_1^2$.

In the presence of macroscopic fluctuations, by definition, the identification between most probable value, average, and observed value of a macroscopic quantity is lost. This would suggest that an expression such as Eq. (3.1) should be interpreted in purely statistical sense. The situation in the example described in Sect. 3.1, of a gas in a tank, with an initial density unbalance in a region V_a, is sketched in Fig. 3.1: if the initial unbalance is large, say, the density n in V_a is substantially higher than in the surrounding, the resulting expansion will be accompanied by turbulent fluctuations; repeating the experiment with the same initial conditions will result in a different turbulent pattern, and the average density profile $n(\mathbf{x}, t)$ will not be observed, typically, in any of the experiments.

These considerations would suggest that the approach in the former section could not be utilized to describe the evolution of an individual macroscopic fluctuation, but that only the mean profile could be obtained. This is unpleasant, because such purely statistical interpretation does not correspond to our idea of density as an instantaneous macroscopic quantity. At the same time, due to the impossibility to apply a mean-field approximation, the resulting equations would become much more difficult to solve.

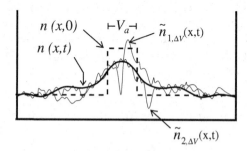

Fig. 3.1 Sketch of the density profile that would be generated, in turbulent conditions, starting from an initial condition $n(\mathbf{x}, 0)$ (*dashed line*). Indicated in figure: $n(\mathbf{x}, t)$ mean profile at time t (*continuous heavy line*); $\tilde{n}_{1,2;\Delta V}(\mathbf{x}, t)$ snapshots of the instantaneous density profile, coarse grained at scale $\Delta V \gg n^{-1}$, in two different experiments starting from the same initial condition $n(\mathbf{x}, 0)$ (*continuous thin lines*)

But what if we insisted on applying a mean-field approximation?

We shall see that, under appropriate conditions, the result could be still perfectly physical, although it would not represent any more an average. In fact, it would describe a possible realization of the fluctuating density profile. This will allow us to recover our intuitive idea of density as a deterministic macroscopic quantity.

A characteristic of systems in turbulent conditions, is sensitive dependence on initial conditions: any small change in the initial conditions (possibly even at the level of the microscopic state Γ), will lead to a different fluctuation pattern in the system. The statistical properties of the fluctuations, though, will remain unaffected. Even more strongly, we could continue to perturb slightly the system, and the fluctuation statistics would remain the same. To focus on the macroscopic component of the fluctuations, let us coarse-grain the fluctuating density at a scale ΔV, much smaller than that of variation of the turbulent fluctuations, but large enough for the thermodynamic limit to be satisfied. Let us now take as initial condition for the coarse-grained mean-field density $n(\mathbf{x}, t)$: $n(\mathbf{x}, 0) = \tilde{n}_{\Delta V}(\mathbf{x}, 0)$, and let it evolve for a short time Δt. At time Δt, the microscopic fluctuations will have started feeding in the nonlinear dynamics, and a macroscopic fluctuating component $\tilde{n}_{\Delta V}(\mathbf{x}, \Delta t) - n(\mathbf{x}, \Delta t)$ would start to be present. If we impose the mean-approximation, however, only that particular evolution, in which microscopic fluctuations have been artificially switched off, will survive and contribute to $n(\mathbf{x}, \Delta t)$. Repeating the procedure many times, will result in a particular evolution of the field $\tilde{n}_{\Delta V}(\mathbf{x}, t)$, that is "special", in that it corresponds to a particular sequence of microscopic fluctuation contributions (the one in which they are all zero), but is still expected to be representative of the statistics of the system.

This suggests the possibility of a definition of quantities, such as the density $n(\mathbf{x}, t)$, or the distribution function $f(\mathbf{y}, t)$, based on coarse-graining of the instantaneous fields $\tilde{n}(\mathbf{x}, t)$ and $\tilde{f}(\mathbf{y}, t)$, instead of averages over realizations, and on disregard of the effect of microscopic fluctuations on their evolution.

We shall stick in the kinetic theory derivation which follows, on this interpretation of the concepts of density and distribution function. Of course, the statistical and the coarse-graining based approach will become equivalent, if the effect of fluctuations on the macroscopic dynamics is negligible.

3.4 Entropy in Kinetic Theory

We have seen in Sect. 2.4, and later in Sect. 3.2, that the Shannon entropy $S_1 \equiv S[\rho_1 \delta y]$ has one of the properties expected in a thermodynamic entropy: growth in correspondence to relaxation to equilibrium. This is a reflection of the fact that relaxation to equilibrium can be seen as a process of homogenization of quantities such as the density and the temperature in the system. A definition of entropy, based on one-molecule properties of the system, however, remains unsatisfactory. A thermodynamic entropy should describe the system as a whole, and non just one of its microscopic parts. The concept of distribution over the Γ-space allows us to make this step. We can define in fact the Shannon entropy associated with the Γ-space PDF ρ_N:

$$S_N(t) = - \int d\Gamma \, \rho_N(\Gamma, t) \ln(\rho_N(\Gamma, t)\delta\Gamma), \qquad (3.17)$$

where $\delta\Gamma = \delta^N y \equiv \delta y_1 \ldots \delta y_N$ fixes the partition of the Γ-space; in the case of a monoatomic gas, of course: $\delta y_k \equiv \delta^3 x_k \delta^3 v_k$, with all the δy_k taken equal: $\delta y_k \equiv \delta y$. In the case of a weakly interacting gas, in which the molecules can be approximated as independent, ρ_N factorizes, $\rho_N(\Gamma, t) \simeq \prod_k \rho_1(y_k, t)$, and S_N becomes a sum of individual molecular contributions:

$$S_N(t) \simeq N S_1(t), \qquad (3.18)$$

where

$$S_1(t) = - \int d^3x d^3v \rho_1(\mathbf{x}, \mathbf{v}; t) \ln(\rho_1(\mathbf{x}, \mathbf{v}; t)\delta y). \qquad (3.19)$$

Equations (3.18) and (3.19) would suggest that the entropy S_N is additive: the entropy of a system composed of weakly interacting macroscopic parts is the sum of the components entropies. Unfortunately, as we shall see below, this is not the case, the reason being the intrinsic microscopic nature of S_N, which, in turn, is a result of the subtle interplay between N and ρ_1 in its definition. This has the consequence that S_N is not well defined on the macroscopic states of a system. The following example illustrates the situation.

3.4.1 Gibbs Paradox and Other Trouble

Consider two volumes $V_{a,b}$ of a certain ideal gas. The two volumes are separated by an infinitely thin wall that prevents molecules from migrating from one volume to the other. Indicate by N_a and N_b, the number of molecules in the two volumes, assume thermodynamic equilibrium conditions with identical values of the temperature and the pressure in both V_a and V_b. This means that the one-molecule velocity PDF will be identical in the two cases, while $\rho_{1a}(\mathbf{x}) = V_a^{-1}$ and $\rho_{2b}(\mathbf{x}) = V_b^{-1}$ (the density is uniform at equilibrium). Hence, from Eqs. (3.17) and (3.18):

$$S_{N_a,a} = N_a[\ln V_a + \tilde{S}], \qquad S_{N_b,b} = N_b[\ln V_b + \tilde{S}], \qquad (3.20)$$

where $\tilde{S} = - \int d^3v \rho_1(\mathbf{v}) \ln(\rho_1(\mathbf{v})\delta y)$ is the contribution to entropy from the molecule velocities. The two systems a and b are by construction totally independent; the PDF on the Γ-space of the composed system factorizes then in the product of the PDF's of the two systems, and the total entropy will be the sum of the entropies $S_N = S_{N_a,a} + S_{N_b,b}$, with $N = N_a + N_b$.

Consider now the systems without the dividing wall. As the two parts have identical temperature and pressure, they are in thermodynamic equilibrium. Thus, they maintain the same macroscopic condition as in the presence of the wall. If the entropy

S_N had a thermodynamic meaning, it should maintain the same value. But this is not the case. Repeating the same calculation leading to Eq. (3.20) gives us:

$$S_N = N(\ln V + \tilde{S}) \neq S_{N_a,a} + S_{N_b,b},$$

where $V = V_a + V_b$. This is the so-called **Gibbs paradox**: the entropy S_N is able to distinguish the two macroscopically identical configurations. The difference between the two, $S_N - (S_{N_a a} + S_{N_b b}) = N_a \ln(V/V_a) + N_b \ln(V/V_b)$, is the gain of entropy from the fact that each molecule has now a volume V in which it can move. This is a purely microscopic effect, of which S_N is taking account. Macroscopically, the difference $S_N - (S_{N_a a} + S_{N_b b})$ could be interpreted as the entropy of mixing two identical substances, which again is a macroscopically meaningless concept.

To understand where we have done something wrong, we must go back to the choice of partition of the sample space (in our case Γ), that we have made at the moment of defining a Shannon entropy S_N. In the case of S_1, the only condition on the partition was uniformity (the volumes $\delta y \equiv \delta^3 x \delta^3 v$ had to be equal), and the fact that δy be larger than any microscale of the system. (The second aspect is rather delicate: if, say, the microscale were discrete, it would be difficult to understand the meaning of a partition that associates multiple cells to a discrete microscopic state of the system).

It turns out that, subdividing the Γ-space in hypercubes of size $\delta\Gamma$, we have done something conceptually similar to considering a partition that is finer than the microscale. Cells in Γ-space identified by coordinates $\Gamma = (\mathbf{y}_1, \mathbf{y}_2, \mathbf{y}_3, \ldots, \mathbf{y}_N)$ and $\Gamma' = (\mathbf{y}_2, \mathbf{y}_1, \mathbf{y}_3, \ldots, \mathbf{y}_N)$, would differ in fact only by exchange of the states of particles 1 and 2. If the two particles are identical, the two states will be physically indistinguishable. A partition that considers Γ and Γ' as separate, would have therefore the same meaning as trying to subdivide the discrete elements of a discrete sample space.

In order to obtain an appropriate partition, we must paste together all the elementary cells, that differ only by a permutations of the $\mathbf{y}_k$'s. Now, in the case of a classical gas, for which the limit $\delta y \to 0$ can be formally taken (see Sect. 3.4.2 below), each cell δy will be occupied, typically, at most by a single molecule, which means that all the component $\mathbf{y}_k$ in the vector $\Gamma = (\mathbf{y}_1, \mathbf{y}_2, \ldots, \mathbf{y}_N)$ will be typically different. Hence the number of cells $\delta\Gamma$ differing by a permutation of $\mathbf{y}_k$ will be $N!$.

The new Shannon entropy, calculated utilizing this partition, with cells $\delta\bar{\Gamma} = \sum_{perm.} \delta\Gamma \simeq N! \delta\Gamma$, will read

$$S(t) \equiv S[\rho_N \delta\bar{\Gamma}] = -\int d\Gamma \, \rho_N(\Gamma, t) \ln(\rho_N(\Gamma, t) N! \delta\Gamma). \tag{3.21}$$

This way of considering cells that differ only because of molecules permutation, as a single indivisible cell of Γ-space, goes under the name of **Boltzmann counting** of the microscopic states of the system. (It must be pointed out that this way of counting states is valid only for dilute gases, in which quantum exchange effects are negligible. We shall return to this point in Sect. 5.3.1).

In the case of independent molecules, we find, from Eq. (3.21)

$$S(t) \simeq NS_1 - \ln N!. \tag{3.22}$$

Since the number N is very large, we can utilize the Stirling approximation $\ln N! \simeq N(\ln N - 1)$. Hence, utilizing Eqs. (3.19) and (3.22), together with the definition $f_1 = N\rho_1$, we obtain

$$S(t) = -\int dy\, f_1(\mathbf{y}, t) \ln(f_1(\mathbf{y}, t)\delta y), \tag{3.23}$$

which is now a quantity defined in terms of the purely macroscopic quantity f_1. This is the expression of entropy, expressed in terms of the distribution f_1, that is going to be utilized in kinetic theory.

Let us verify that the new definition cures the Gibbs paradox.

Consider again the system formed by two volumes of gas, V_a and V_b, at equal temperature and pressure. In thermodynamic equilibrium, we will have $f_1(\mathbf{x}, \mathbf{v}) = N\rho_1(\mathbf{x}, \mathbf{v}) = n\rho_1(\mathbf{v})$, where $n = N_a/V_a = N_b/V_b$ is the density of the gas, equal in the two volumes. Hence, from Eq. (3.23), we have, in place of Eq. (3.20):

$$S_{a,b} = N_{a,b}[\tilde{S} - \ln n]. \tag{3.24}$$

Similarly for the entropy of the system, obtained putting together V_a and V_b: $S = N[\tilde{S} - \ln n]$ We thus find $S = S_a + S_b$. The fact that entropy now depends on the density, and not on the volume of the gas, guarantees that the additivity property is satisfied.

As we have done with other macroscopic quantities, such as n, and f, we can now introduce an instantaneous version of the entropy, that is the definition of entropy adopted by Boltzmann[1]:

$$\tilde{S}_{\Delta y}(t) = -\int dy\, \tilde{f}_{1,\Delta y}(\mathbf{y}, t) \ln(\tilde{f}_{1,\Delta y}(\mathbf{y}, t)\delta y), \tag{3.25}$$

(we have to consider a coarse-grained version of $\tilde{f}_1$, to avoid problems of Dirac deltas in the argument of the logarithm). Contrary to S, the quantity $\tilde{S}_{\Delta y}$ will contain fluctuations, even at thermodynamic equilibrium. These fluctuations will be negligible only in the thermodynamic limit $\Delta N = \tilde{f}_{1,\Delta y}\Delta y \gg 1$.

3.4.2 Quantum Effects

As we have said, it is necessary that the partition of Γ-space be not so fine to treat as distinct, physically identical states. This has forced us to glue together hypercubes $\delta\Gamma$ that differ only by particle permutations. At the same time, we must be careful that the individual hypercubes are not too small that the physics at the cell scale

---

[1] More precisely, the definition by Ludwig Boltzmann was $H = -\tilde{S}$, from which the name H-theorem.

changes. In particular, cells should not be so small for quantum effects to become important. This scale is fixed by the Planck constant $\hbar = 6.6 \times 10^{-27}$erg $\cdot$ s.

Now, while particle indistinguishability affected the geometric structure of the partition at macroscopic scale (the scale of the physical particle separation), to disregard quantum effects, it seems that it would be enough to consider cells δy, that, although microscopic, are not *too* small. Unfortunately, also in this case, "not too small" may still mean "macroscopic".

Cells of one-molecule phase space, with minimal size $\delta y \sim (\hbar/m)^3$, with m the molecular mass, correspond basically to a one-molecule quantum state. Quantum exchange effects occur, when more than one molecule try to occupy the same quantum state; an example of such an effect is provided by the Pauli exclusion principle: contrary to the case of classical particles, it is impossible for two fermions to occupy the same cell δy. We can interpret classically this result as a sort of repulsion between particles, whence the name of quantum exchange "interaction". In order for such interaction to be considered negligible, it is necessary that the probability of situations, in which more than one particle try to occupy a same cell δy, is small.

We stress the importance of this point, as the possibility of factorizing the Γ-space PDF, $\rho_N(\Gamma, t) \simeq \prod_k \rho_1(\mathbf{y}_k, t)$, that we have assumed implicitly from the start, rests also on the possibility of neglecting quantum mechanical exchange interactions.

The condition to avoid exchange interactions is that there are many more cells available than particles. The number of cells can be estimated as the ratio of the volume of phase space, available to each molecule, to the cell size $(\hbar/m)^3$. The phase space volume is the product of the physical volume V, and the volume in velocity space, which, in the case of a classical ideal gas, can be estimated from the thermal velocity $v_{th} = (KT/m)^{1/2}$, with T the temperature. The phase space volume is therefore $V(mv_{th})^3$, and the number of cells is $G = V(mv_{th}/\hbar)^3$. The condition $G \gg N$ is therefore, indicating $n = N/V$:

$$n(KT/m)^{-3/2} \ll \hbar^{-3}. \qquad (3.26)$$

In order for a gas to behave classically, it is necessary that it is dilute, and that it is not too cold: this is the condition for Eqs. (3.18) and (3.19) to be applicable. We shall focus in these notes on classical systems for which the condition in Eq. (3.26) is satisfied, which is tantamount to stating that the limit $\hbar \to 0$, and therefore, also $\delta y \to 0$, can be safely taken.

3.5 The Boltzmann Equation

Let us pass to analysis of the real problem: the dynamics of a dilute gas of molecules, that interact by collisions. We want to derive an evolution equation for the distribution $f_1(\mathbf{x}, \mathbf{v}; t)$, analogous to the one obtained in the case of the interacting random walkers of Sect. 3.2. The main difference is that now, each component of the system is a molecule described by a six-dimensional phase space labeled by the coordinates $\mathbf{x}$ and $\mathbf{v}$.

We can write the equation of motion for a molecule, isolating the contribution from the collisions, in the form:

$$\dot{\mathbf{x}}_k = \mathbf{v}_k, \quad m\dot{\mathbf{v}}_k = \mathbf{F}(\mathbf{x}_k, \mathbf{v}_k; t) = [\mathbf{F}^{ext}(\mathbf{x}_k, \mathbf{v}_k; t) + \mathbf{F}^{coll}(\{\mathbf{x}_i, \mathbf{v}_i; i = 1, \ldots N\}; t)].$$
(3.27)

Notice the dependence of the collisional force $\mathbf{F}^{coll}$, on the position and velocities of all the molecules in the gas, which reveals the interaction nature of this term. We keep the velocity dependence in $\mathbf{F}$, to account for the possibility of Lorentz forces. In the absence of collisions, the equation of motion (3.27) would be in the form $\dot{\mathbf{y}} = \mathbf{V}(\mathbf{y}, t)$, where $\mathbf{y} = (\mathbf{x}, \mathbf{v})$, $\mathbf{V} = (\mathbf{v}, \mathbf{F}^{ext}/m)$, and the evolution equation for f_1 would be the continuity equation

$$\frac{\partial f_1}{\partial t} + \nabla_{\mathbf{y}} \cdot (f_1 \mathbf{V}) = 0.$$
(3.28)

The collisional force $\mathbf{F}^{coll}$ adds a contribution to $\partial f_1/\partial t$, that we indicate as $(\partial f_1/\partial t)_{coll}$, so that Eq. (3.28) takes the form

$$\frac{\partial f_1}{\partial t} + \mathbf{v} \cdot \nabla_{\mathbf{x}} f_1 + \frac{1}{m} \nabla_{\mathbf{v}} \cdot (F^{ext} f_1) = \left(\frac{\partial f_1}{\partial t}\right)_{coll}.$$
(3.29)

The fundamental hypothesis we make, to determine the form of the term $(\partial f/\partial t)_{coll}$, is the absence of memory in the collision process: the final velocities of the molecules participating in a collision, depend only on their velocity before, independently on their previous collision histories. We can therefore represent the velocity of the molecules, as a Markov process. This underlies a statistical description of the collision process, that can be seen as the result of a choice of microscale δx, above that of the molecules and their interaction length. The only information that we utilize, will be that of binary, elastic and instantaneous collisions.

The structure of the term $(\partial f_1/\partial t)_{coll}$ can be determined in analogy with what we have done in the case of the toy model of Sect. 3.2. Let us indicate by $w(\mathbf{v}'_{1,2} \rightarrow \mathbf{v}_{1,2})$ the rate at which molecules at position $\mathbf{x}_1 = \mathbf{x}_2$ make transition, due to collisions, from initial velocities $\mathbf{v}'_1$ and $\mathbf{v}'_2$ to $\mathbf{v}_1$ and $\mathbf{v}_2$. The evolution equation will take the form of a master equation, in which the effect of the collision is taken into account through

$$\left(\frac{\partial f_1(\mathbf{x}, \mathbf{v}_1; t)}{\partial t}\right)_{coll} = \int d^3 v'_1 d^3 v'_2 d^3 v_2 \{w(\mathbf{v}'_{1,2} \rightarrow \mathbf{v}_{1,2}) f_2(\mathbf{x}, \mathbf{v}'_{1,2}; t)$$
$$- w(\mathbf{v}_{1,2} \rightarrow \mathbf{v}'_{1,2}) f_2(\mathbf{x}, \mathbf{v}_{1,2}; t)\}.$$

Symmetry of elastic collisions, under space and time inversion, allows us to write $w(\mathbf{v}_{1,2} \rightarrow \mathbf{v}'_{1,2}) = w(\mathbf{v}'_{1,2} \rightarrow \mathbf{v}_{1,2})$, so that

$$\left(\frac{\partial f_1(\mathbf{x}; \mathbf{v}_1; t)}{\partial t}\right)_{coll} = \int d^3 v_2 d^3 v'_1 d^3 v'_2 \, w(\mathbf{v}_{1,2} \rightarrow \mathbf{v}'_{1,2})[f_2(\mathbf{x}; \mathbf{v}'_{1,2}; t) - f_2(\mathbf{x}; \mathbf{v}_{1,2}; t)].$$

Substituting into Eq. (3.29), we obtain an evolution equation for f_1, which suffers, as in the case of the interacting walker problem, of the problem of closure: to solve it, we would need to know the two-molecule distribution f_2. As before, the only practicable solution is to adopt the mean-field approximation $f_2(\mathbf{x}, \mathbf{v}_{1,2}; t) \simeq f_1(\mathbf{x}, \mathbf{v}_1; t)f_1(\mathbf{x}, \mathbf{v}_2; t)$. As discussed in Sect. 3.3, this is consistent with an interpretation of the distribution function f as an instantaneous coarse-grained quantity. The result is the following **Boltzmann equation**:

$$\frac{\partial f_1}{\partial t} + \mathbf{v} \cdot \nabla_{\mathbf{x}} f_1 + \frac{1}{m} \nabla_{\mathbf{v}} \cdot (F^{ext} f_1) = \left(\frac{\partial f_1}{\partial t}\right)_{coll}, \tag{3.30}$$

in which the collision term is given by

$$\left(\frac{\partial f_1(\mathbf{x}, \mathbf{v}_1; t)}{\partial t}\right)_{coll} = \int d^3 v_2 d^3 v_1' d^3 v_2' \; w(\mathbf{v}_{1,2} \to \mathbf{v}_{1,2}')$$
$$\times [f_1(\mathbf{x}; \mathbf{v}_1'; t)f_1(\mathbf{x}; \mathbf{v}_2'; t) - f_1(\mathbf{x}; \mathbf{v}_1; t)f_1(\mathbf{x}; \mathbf{v}_2; t)]. \tag{3.31}$$

The Boltzmann equation has been incredibly successful in describing the dynamics of gases. Thus, this is perhaps the best justification for the approximations adopted to reach Eqs. (3.30) and (3.31). The reason why such approximations work so well, lies probably in the ability of collisions to destroy correlations between molecules. The idea is that molecules 1 and 2 arrive at point $\mathbf{x}$ after a sequence of collisions that have made them essentially uncorrelated (molecular chaos hypothesis). It is easy to see that the molecular chaos hypothesis $f_2(\mathbf{x}, \mathbf{v}_1; \mathbf{x}, \mathbf{v}_2; t) = f_1(\mathbf{x}, \mathbf{v}_1; t)f_1(\mathbf{x}, \mathbf{v}_2; t)$, $\mathbf{v}_1 \neq \mathbf{v}_2$, plays in velocity space, the same role played in real space by the condition Eq. (3.13).

3.5.1 The Maxwell-Boltzmann Distribution

We can see that the Maxwell-Boltzmann distribution

$$f^{MB}(\mathbf{x}, \mathbf{v}) = n_0 \left(\frac{m}{2\pi KT}\right)^{3/2} \exp\left(-\frac{mv^2}{2KT}\right), \tag{3.32}$$

given a uniform profile of density $n(\mathbf{x}, t) = n$, and temperature $T(\mathbf{x}, t) = T$, in the absence of external forces, is a stationary solution of the Boltzmann equation (for lighter notation, we shall drop from now on subscript 1 on the one-molecule distribution f_1). Substituting Eq. (3.32) into Eq. (3.30), gives us:

$$\frac{\partial f^{MB}}{\partial t} = \left(\frac{\partial f^{MB}}{\partial t}\right)_{coll}.$$

To prove stationarity we must verify therefore that $(\partial f / \partial t)_{coll} = 0$.

We start by noting that, in each collision, energy conservation imposes (recall that we are dealing with elastic collisions):

$$v_1^2 + v_2^2 = {v_1'}^2 + {v_2'}^2.$$

Substituting $f = f^{MB}$ into the expression for $(\partial f/\partial t)_{coll}$ given in Eq. (3.31), we find therefore, writing $f^{MB} = Ce^{-\alpha v^2}$, with $C = n_0(m/(2\pi KT))^{3/2}$ and $\alpha = m/(2KT)$:

$$[f^{MB}(\mathbf{x}; \mathbf{v}_1')f^{MB}(\mathbf{x}; \mathbf{v}_2') - f^{MB}(\mathbf{x}; \mathbf{v}_1)f^{MB}(\mathbf{x}; \mathbf{v}_2)]$$
$$= C^2[\exp\{-\alpha({v_1'}^2 + {v_2'}^2)\} - \exp\{-\alpha({v_1}^2 + {v_2}^2)\}] = 0.$$

Thus, $(\partial f^{MB}/\partial t)_{coll} = 0$, as required.

We shall show in Sect. 5.3, that the entropy of a distribution is maximum, for given value of the variance, when its profile is Gaussian. Thus, the Maxwell-Boltzmann distribution, beyond being the stationary distribution for the Boltzmann equation, is also the maximum entropy distribution for the given temperature. The H-theorem, that we shall derive below, guarantees that systems obeying the Boltzmann equation, are characterized by a law of growth for the entropy. We expect therefore, that f^{MB} be the equilibrium distribution as well, on which the system will relax, independently of initial conditions (for given values of the temperature and volume).

3.5.2 Entropy in Velocity Space

The dependence of f on the velocity $\mathbf{v}$, leads to a new component in the entropy. We have seen in Sect. 2.4, that the spatial component of the entropy is maximum in correspondence to a uniform distribution. We shall see in the next section, that the velocity component of the entropy is maximum, for given value of the RMS velocity, in correspondence of the Gaussian distribution, i.e. the Maxwell-Boltzmann distribution Eq. (3.32). We can easily provide an example of non-Maxwellian distribution whose entropy is smaller (for given RMS velocity) than the corresponding Maxwellian: that for two "monochromatic" jets, $\rho_1(\mathbf{v}) = \frac{1}{2}[\delta(\mathbf{v} - \mathbf{v}_0) + \delta(\mathbf{v} + \mathbf{v}_0)]$, with $\langle v \rangle = 0$ and $\sigma_v^2 = v_0^2$. It is easy to be convinced that this is actually the lowest entropy state that realizes the condition $\langle v \rangle = 0$, $\langle v^2 \rangle = v_0^2$. Collisions will break up the jets, leading to an increase of the entropy, and relaxation of ρ_1 on the distribution ρ_1^{MB}, corresponding to $KT = (1/3)mv_0^2$.

Substituting Eq. (3.32) into Eq. (3.23), we obtain the following expression for the entropy:

$$S = N\left(\frac{3}{2}\ln T + \ln(V/N) + \zeta\right), \tag{3.33}$$

that is, to within multiplicative constants, the expression for the entropy of a gas at temperature T, that we know from thermodynamics. The dependence of the entropy on the temperature, already known in thermodynamics, takes here a new meaning: it is associated with the fact that $\langle v^2 \rangle \propto T$, so that higher T will correspond to a more spread-out distribution for v.

3.5.3 The H Theorem

The Boltzmann equation satisfies a growth condition for the entropy $S(t)$ in Eq. (3.23). This condition is valid, whatever the initial condition for the distribution f. This is the essential condition that will allow to make a direct connection with the thermodynamics of the system.

Let us calculate the growth rate $\dot{S}$ explicitly. Proceeding as in the case of the toy model of Sect. 3.2 (see note at page 46), we find immediately

$$\dot{S} = -\int d^3x d^3v \dot{f}(\mathbf{x}, \mathbf{v}; t) \ln f(\mathbf{x}, \mathbf{v}; t).$$

(It is possible to see that the constant δy disappears from all calculations and we thus drop it from the start). We see that the only contribution to $\dot{S}$ is from collisions, i.e. from $(\partial f / \partial t)_{coll}$.

Using the divergence theorem on the term involving $\nabla_{\mathbf{x}}$ in the Boltzmann equation (3.30), and disregarding terms at $\mathbf{x} \to \infty$, we find in fact:

$$-\int d^3x d^3v \ln f \nabla_{\mathbf{x}} \cdot (\mathbf{v}f) = \int d^3x d^3v \nabla_{\mathbf{x}} \cdot (\mathbf{v}f) = 0.$$

In analogous way, for the external force contribution:

$$-\int d^3x d^3v \ln f \nabla_{\mathbf{v}} \cdot (\mathbf{F}^{ext}f) = \int d^3x d^3v \mathbf{F}^{ext} \cdot \nabla_{\mathbf{v}} f = -\int d^3x d^3v f \nabla_{\mathbf{v}} \cdot \mathbf{F}^{ext},$$

that is zero, both in the case of forces that do not depend on the velocity, and in the presence of magnetic fields (in this last case, we can exploit the relation $\nabla_{\mathbf{v}} \cdot [\mathbf{v} \times \mathbf{B}] = \mathbf{B} \cdot [\nabla_{\mathbf{v}} \times \mathbf{v}] - \mathbf{v} \cdot [\nabla_{\mathbf{v}} \times \mathbf{B}] = 0$).

We find in the end, using Eq. (3.31):

$$\dot{S} = -\int d^3x d^3v (\partial f/\partial t)_{coll} \ln f(\mathbf{x}, \mathbf{v}; t)$$

$$= \int d^3x \int d\mu_{\mathbf{v}} w(\mathbf{v}'_{1,2} \to \mathbf{v}_{1,2})[f(1')f(2') - f(1)f(2)] \ln f(1),$$

where the shorthands $d\mu_v = d^3v_1 d^3v_2 d^3v'_1 d^3v'_2, f(1) \equiv f(\mathbf{x}, \mathbf{v}_1; t)$, and so on, have been adopted. We see at once that, thanks to the symmetry of $d\mu_v w[\dots]$, the result of the integration will not change if we substitute $\ln f(1)$ with $\ln f(2)$ in the integral. We can write therefore:

$$\dot{S} = -\frac{1}{2} \int d^3x \int d\mu_v w(\mathbf{v}'_{1,2} \rightarrow \mathbf{v}_{1,2})[f(1')f(2') - f(1)f(2)] \ln(f(1)f(2)).$$

Exploiting symmetry under space and time inversion, $w(\mathbf{v}'_{1,2} \rightarrow \mathbf{v}_{1,2}) = w(-\mathbf{v}_{1,2} \rightarrow -\mathbf{v}'_{1,2}) = w(\mathbf{v}_{1,2} \rightarrow \mathbf{v}'_{1,2})$, allows us to write:

$$\dot{S} = -\frac{1}{2} \int d^3x \int d\mu_v w(\mathbf{v}'_{1,2} \rightarrow \mathbf{v}_{1,2})[f(1)f(2) - f(1')f(2')] \ln(f(1')f(2')).$$

Summing the two expressions, we find:

$$\dot{S} = \frac{1}{4} \int d^3x \int d\mu_v w(\mathbf{v}'_{1,2} \rightarrow \mathbf{v}_{1,2})[f(1')f(2') - f(1)f(2)]$$
$$\times [\ln(f(1')f(2')) - \ln(f(1)f(2))], \tag{3.34}$$

and the integrand will always be positive. We thus get the entropy growth condition

$$\dot{S} \geq 0,$$

where equality is satisfied only if the distribution is Maxwellian. We have not demonstrated yet that the distribution has to be spatially homogeneous. Equation (3.34) tells us in fact that $\dot{S} = 0$, also for a spatially inhomogeneous Maxwellian f. As we shall see in the coming sections, however, any spatial dependence in a Maxwellian distribution, will lead quickly to a non-Maxwellian component in f, and therefore to $\dot{S} > 0$. It can actually be said that thermodynamic equilibrium is achieved only thanks to the development of a non-Maxwellian component in the distribution f.

As a last comment, we point out that the H-theorem statement, that entropy is a monotonously increasing function of time, does not apply to the instantaneous entropy $\tilde{S}$. The instantaneous entropy contains in fact a fluctuating component, of which S is free.[2] (The mean-field approximation, leading Eq. (3.31), destroys all fluctuation contributions to $(\partial f / \partial t)_{coll}$ and consequently to $\dot{S}$). This provides a first illustration of the nature of fluctuations, as departures from equilibrium, that temporarily lead to a decrease of entropy in the system.

---

[2] We speak here of microscopic fluctuations. In the presence of macroscopic fluctuations, the mean-field approximation leading to the Boltzmann equation, implies that f represents a realization of the macroscopic fluctuations (see Sect. 3.3).

3.6 The Fluid Limit

The approach to thermodynamic equilibrium for a dilute gas, that is initially in a generic non-Maxwellian state, takes place in substantially two steps. First, f will relax to a locally Maxwellian form:

$$f^{MB}(\mathbf{x}, \mathbf{v}; t) = n(\mathbf{x}, t)\left(\frac{m}{2\pi KT(\mathbf{x}, t)}\right)^{3/2} \exp\left(-\frac{m|\mathbf{v} - \mathbf{u}(\mathbf{x}, t)|^2}{2KT(\mathbf{x}, t)}\right), \qquad (3.35)$$

in which we admit the possibility of a mean flow $\mathbf{u}(\mathbf{x}, t)$. This condition is typically referred to as **kinetic equilibrium**.

The relaxation time to kinetic equilibrium is usually very short. Due to the violent nature of molecular collisions, and their ability to destroy any information on the state of the molecules involved, just a few collisions are in fact sufficient. The time required to reach thermal equilibrium, i.e. e state in which profiles for n, $\mathbf{u}$ and T, are uniform, is much longer, and is determined by the pressure and the viscosity forces, as well as by the effect of heat diffusion.

We can estimate the collision time τ_{coll}, modeling the molecule interaction as if the molecules were billiard balls. The parameters that describe the system will be therefore just the density n, the molecular radius a (that fixes the interaction distance for the molecules) and the thermal velocity v_{th} (that determines the typical collision velocity for the molecules).

We obtain immediately the dimensional estimates for the mean free path λ and the collision time τ_{coll} of the molecules:

$$\lambda \sim \frac{1}{na^2}, \qquad \tau_{coll} \sim \frac{\lambda}{v_{th}}. \qquad (3.36)$$

Note The expression of λ can be obtained from geometric considerations, from the observation that the typical separation between molecules is $l \sim n^{-1/3}$. In every molecular volume n^{-1}, there will be typically one molecule. The probability for a molecule crossing a certain molecular volume, to undergo a collision in it, will be $P \sim (a/l)^2$. The number of molecular volumes that molecule has to cross, before making a hit, will be therefore $\sim P^{-1} \sim (l/a)^2$, and the distance crossed during that time will be $\lambda \sim l/P \sim l^3/a^2 = 1/(na^2)$.∎

As we have said, the collision time τ_{coll} is typically extremely short on the scale of the thermodynamic processes. To make an example, consider a gas in typical atmospheric conditions:

$$n \sim 10^{20} \text{cm}^{-3}, \quad a \sim 10^{-8} \text{cm}, \quad v_{th} \sim 3 \cdot 10^5 \text{cm/s};$$

from (3.36) we would get:

$$\lambda \sim 10^{-4} \text{cm}, \qquad \tau_{coll} \sim 10^{-9} \text{s},$$

which, for the system in consideration, is indeed extremely short.

Fig. 3.2 Mechanism of generation of a the non-Maxwellian
component $g = f - f^{MB}$ in
presence of a temperature
gradient

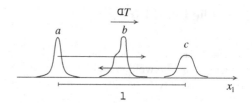

Once a kinetic equilibrium condition $f = f^{MB}(\mathbf{x}, \mathbf{v}; t)$ has been achieved, from
Eq. (3.35), the problem is reduced from that of determining $f(\mathbf{x}, \mathbf{v}; t)$, to the much
simpler one of determining the evolution of the fields $n(\mathbf{x}, t)$, $\mathbf{u}(\mathbf{x}, t)$ and $T(\mathbf{x}, t)$. This
could be done e.g. substituting Eq. (3.35) into Eq. (3.30), and exploiting the property
$(\partial f^{MB}/\partial t)_{coll} = 0$, to convert the Boltzmann equation (3.30) into a set of PDE's for
n, $\mathbf{u}$ and T. This is the moment equation technique outlined at the end of Sect. 2.7.1.

Unfortunately, as discussed in Sect. 3.5.3, entropy could not grow if the distribution were fully Maxwellian. In order for the fields n, $\mathbf{u}$ and T to relax to their equilibrium value, a spatially non-uniform distribution f must have a non-Maxwellian
component $g = f - f^{MB}$. The mechanism leading to the generation of this component is illustrated in Fig. 3.2, in the case of a temperature gradient in the gas,
directed along x_1. Molecules crossing b, will have originated from collisions at a
and c, where they thermalized at temperature T_a and T_c, with $T_a < T_c$. When they
cross b, the molecules maintain the same distribution (more precisely, the right side
for those coming from a; the left side from those coming from c). The end result is a
distribution that would be non-Maxwellian even when the original profiles in a and
b were Maxwellian.

To estimate the magnitude of the non-Maxwellian component, we make the smallness assumption $g \ll f^{MB}$ (in some appropriate norm) and linearize the collisional
term Eq. (3.31):

$$\left(\frac{\partial f}{\partial t}\right)_{coll} \sim -\frac{f - f^{MB}}{\tau_{coll}}. \tag{3.37}$$

This is called the BGK approximation for the collisional term, and implies an exponential relaxation of f to f^{MB} at time scale τ_{coll}. We can estimate g, in the absence of
external forces, substituting Eq. (3.37) directly into the Boltzmann equation (3.30):

$$\left(\frac{\partial}{\partial t} + \mathbf{v} \cdot \nabla_x\right)f \sim -\frac{g}{\tau_{coll}}.$$

Approximating $f \simeq f^{MB}$ in the LHS of this equation, and indicating by L the spatial
scale of variation of f^{MB}, we obtain the estimate:

$$g \sim \frac{v_{th}\tau_{coll}f^{MB}}{L} \sim \frac{\lambda}{L}f^{MB}. \tag{3.38}$$

Thus, the correction g will be small if f^{MB} varies little on the scale of the mean free
path λ.

The $L \gg \lambda$ regime is called the **fluid limit** for the gas. In this limit, the fields n, $\mathbf{u}$ and T provide a complete description, that corresponds to our experience of a gas as a fluid whose physical state is determined by the three "fluid quantities": density, fluid velocity and temperature. The condition $L \gg \lambda$ can be interpreted as a condition of **local thermodynamic equilibrium** at scales l, with $\lambda \ll l \ll L$. In fact, in order to have thermodynamic equilibrium, collisions must be able to homogenize the parameters n, $\mathbf{u}$ and T in the gas. The volume in which homogenization takes place, must be much larger than λ, otherwise, different portions of the volume would be unable to communicate through collisions. In other words, λ is the shortest scale at which collisions could smear out inhomogeneities in n, $\mathbf{u}$ and T, and a local thermodynamic equilibrium could be realized.

3.7 Thermodynamic Meaning of Temperature and Entropy

Relaxation to thermodynamic equilibrium of a gas already in kinetic equilibrium, is achieved through homogenization of the fluid quantities n, $\mathbf{u}$ and T. In the process, the entropy of the system will rise to its maximum, given by Eq. (3.33). The entropy of a gas, whose distribution is locally Maxwellian, as described by Eq. (3.35), is in the form:

$$S(t) = \int_V d^3x \, n(\mathbf{x}, t)\left(\frac{3}{2} \ln T(\mathbf{x}, t) - \ln n(\mathbf{x}, t) + \zeta\right). \tag{3.39}$$

We know from the analysis in Sect. 2.4 that a spatially uniform distribution $n(\mathbf{x}, t) = \bar{n}$ will correspond to a maximum of S. We can show that the same occurs when the temperature becomes uniform in the domain. This will complete the connection between the concept of statistical entropy, with which we have been working so far, and that of thermodynamic entropy. At the same time, we will be able to give a thermodynamic meaning to the concept of temperature, that too has had for us, up to this point, only a statistical meaning.

Let us divide our volume V in two parts A and B. Suppose the gas in the two parts, in thermal equilibrium, and suppose equal values of the density $n_A = n_B$. We want to verify that global thermal equilibrium corresponds to equal temperatures, $T_A = T_B$. Suppose the volume V isolated, so that the total internal energy $E = E_A + E_B$ is conserved.

From the additive nature of entropy, we shall have at equilibrium:

$$S = S_A(E_A) + S_B(E - E_A).$$

The maximum entropy condition, for constant E, can be expressed as a condition of maximum with respect to the energy of one of the parts, say E_A:

$$\frac{dS}{dE_A} = 0 \Rightarrow \frac{dS_A}{dE_A} = \frac{dS_B}{dE_B}; \qquad \frac{d^2 S_A}{dE_A^2} + \frac{d^2 S_B}{dE_B^2} < 0. \tag{3.40}$$

In the case of a gas in thermal equilibrium, the total energy in E coincides with the thermal energy:

$$E = \frac{3}{2}NKT,$$ (3.41)

which, substituting together with Eq. (3.33), into Eq. (3.40), gives:

$$\frac{dS_A}{dE_A} = \frac{dS_A}{dT_A}\left(\frac{dE_A}{dT_A}\right)^{-1} = \frac{1}{KT_A}.$$ (3.42)

Similarly for S_B, that gives $T_A = T_B$, as expected. Substituting again, with Eq. (3.41), into Eq. (3.42), we easily verify that we really have a maximum:

$$\frac{d^2 S_A}{dE_A^2} = \frac{2N}{3}\frac{d}{dE_A}\frac{1}{E_A} = -\frac{2N}{3E_A^2} < 0.$$

We thus recover the standard thermodynamic notion of temperature, as that parameter whose equality among systems tells us that they are in equilibrium. If one of the systems has higher temperature, heat must flow to the other to recover equilibrium.

The property of entropy maximum at equilibrium, has a geometric interpretation in the case the two subsystems A and B are identical. As in the case of the Shannon entropy, a central role is played by convexity of the function. In this case, the two entropies S_A and S_B have identical functional form, and the values of S can be determined studying a single curve, as illustrated in Fig. 3.3. We see in figure that the points of the curve lie below the tangent. This implies $S_A(E_A) + S_B(E_B) \leq S(E/2)$, with equality satisfied at equilibrium, $E_{A,B} = \bar{E}_{A,B} = E/2$ (the tangency point). We thus see that the condition that entropy be maximum at equilibrium, is allowed by the fact that the entropy is a convex increasing function of the energy (compare with the discussion at the end of Sect. 2.2). On the other hand, we have seen that such properties are associated with the fact that temperature is positive and an increasing function of the energy.

We can recover at this point the connection between statistical entropy and thermodynamic entropy. Imagine that A and B are separated by a heat conducting wall, and suppose that there was initially a temperature difference between the two parts.

Fig. 3.3 Entropy profile in the subsystems as function of their internal energy. At equilibrium $S(E) = 2S_{A,B}(E/2) = 2\bar{S}$. The slope of the tangent in $E_{A,B} = \bar{E}_{A,B} = E/2$ equals the inverse of the equilibrium temperature T

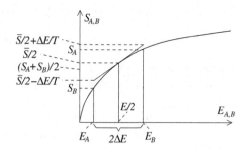

The energy of the two parts will change during relaxation to equilibrium, and the entropy will change in consequence. If the initial temperature gap is small, we can assume that the two parts remain internally in equilibrium during the process, so that Eq. (3.33) continues to apply in the two cases. The entropy change can then be obtained inverting Eq. (3.42):

$$dS_A = \frac{dE_A}{KT_A}, \tag{3.43}$$

where dE_A is just the heat $d\!\!{}^-Q$ received by A.[3] We thus recover (to within a constant factor K) the standard definition of thermodynamic entropy $dS = d\!\!{}^-Q/T$, for a body that exchanges heat (reversibly) with the environment.

3.8 The Equations of Fluid Mechanics

We have seen that a proper description of the behavior of a non-equilibrium gas, requires taking into account the existence of a non-Maxwellian component in the distribution f. This means that we cannot derive the evolution equations for the fluid quantities n, $\mathbf{u}$ and T, by hypothesizing a strictly Maxwellian form for the distribution f, locally, and substituting it in the Boltzmann equation (3.30).

In order to take into account the contribution by the non-Maxwellian component of f, to the gas dynamics, it is necessary to define n, $\mathbf{u}$ and T, independently of the specific form of the distribution f:

$$n(\mathbf{x}, t) = \int d^3 v f(\mathbf{x}, \mathbf{v}; t); \qquad \mathbf{u}(\mathbf{x}, t) = \frac{1}{n(\mathbf{x}, t)} \int d^3 v \mathbf{v} f(\mathbf{x}, \mathbf{v}; t);$$

$$T(\mathbf{x}, t) = \frac{m}{3Kn(\mathbf{x}, t)} \int d^3 v |\mathbf{v} - \mathbf{u}(\mathbf{x}, t)|^2 f(\mathbf{x}, \mathbf{v}, t). \tag{3.44}$$

In this way, the fluid quantities can be interpreted as the first moments, with respect to v, of the distribution function f, or, alternatively, as the conditional averages with respect to the one-particle distribution $\rho_1(\mathbf{x}, \mathbf{v}, t)$: $\mathbf{u}(\mathbf{x}, t) = \langle \mathbf{v} \mid X; t \rangle$, $T(X, t) = \frac{m}{3k} \langle |v - u(X, t)|^2 | X, t \rangle$. Their evolution equations could be obtained taking the respective moments of the Boltzmann equation (3.30):

$$\int d^3 v I_k(\mathbf{v}) \left[\frac{\partial f}{\partial t} + \mathbf{v} \cdot \nabla_x f + \frac{1}{m} \nabla_\mathbf{v} \cdot (\mathbf{F}^{ext} f) \right] = \int d^3 v I_k(\mathbf{v}) \left(\frac{\partial f}{\partial t} \right)_{coll}, \tag{3.45}$$

where $I_0(\mathbf{v}) = 1$, $I_1(\mathbf{v}) = m\mathbf{v}$ and $I_2(\mathbf{v}) = \frac{mv^2}{2}$.

---

[3] We recall the notation $d\!\!{}^-$ to indicate that the differential is not exact: while dS_A is the differential of a function $S_A(T_A, V_A)$, there is no function $Q_A(T_A, V_A)$. See Sect. 4.3.

Let us carry out the calculation explicitly for $k = 0$. We find at once:

$$\frac{\partial}{\partial t} \int d^3 v f + \nabla_{\mathbf{x}} \cdot \int d^3 v f \mathbf{v} + \frac{1}{m} \int d^3 v \nabla_{\mathbf{v}} \cdot (\mathbf{F}^{ext} f) = \int d^3 v \left(\frac{\partial f}{\partial t}\right)_{coll}.$$

Using the divergence theorem, the third integral to LHS of the equation becomes a surface integral at $\mathbf{v} \to \infty$, that could be put equal to zero (in order for Eq. (3.45) to make sense, it is necessary that $f \to 0$ for large $\mathbf{v}$ at least as v^{-6}). We are left therefore with the equation

$$\frac{\partial n}{\partial t} + \nabla_{\mathbf{x}} \cdot (n\mathbf{u}) = \int d^3 v \left(\frac{\partial f}{\partial t}\right)_{coll}.$$

We can see that the RHS of this equation is identically zero, thanks to particle conservation in the collision process. We have indeed, exploiting the symmetry $w(\mathbf{v}'_{1,2} \to \mathbf{v}_{1,2}) = w(\mathbf{v}_{1,2} \to \mathbf{v}'_{1,2})$, and indicating again $d\mu_v \equiv d^3 v_1 d^3 v_2 d^3 v'_1 d^3 v'_2$, $f(1) \equiv f(\mathbf{x}, \mathbf{v}_1)$, etc:

$$\int d^3 v_1 \left(\frac{\partial f(\mathbf{x}, \mathbf{v}_1; t)}{\partial t}\right)_{coll} = \int d\mu_v w(\mathbf{v}_{1,2} \to \mathbf{v}'_{1,2})[\rho_1(1')\rho_1(2') - \rho_1(1)\rho_1(2)] = 0.$$

We thus obtain the continuity equation:

$$\frac{\partial n}{\partial t} + \nabla_{\mathbf{x}} \cdot (n\mathbf{u}) = 0. \tag{3.46}$$

This result could be expected: the Boltzmann equation is itself a continuity equation with respect to $(\mathbf{x}, \mathbf{v})$, forced by the collision term $(\partial f / \partial t)_{coll}$. It is not a surprise, therefore, that integrating over $\mathbf{v}$ leads to a continuity equation with respect to $\mathbf{x}$; the fact that $(\partial f / \partial t)_{coll} = 0$, reflects simply the fact that collisions do not produce or destroy molecules.

A procedure similar to the one leading to Eq. (3.46) can be applied to $I_k(\mathbf{v})$, for $k > 0$. The procedure is detailed in the Appendix. We find for $k = 1$ and $k = 2$:

$$nm\left(\frac{\partial}{\partial t} + \mathbf{u} \cdot \nabla_{\mathbf{x}}\right) u_i = -\frac{\partial P}{\partial x_i} + \mathscr{F}_i^{ext} - \frac{\partial \pi_{ij}}{\partial x_j}, \tag{3.47}$$

and

$$n\left(\frac{\partial}{\partial t} + \mathbf{u} \cdot \nabla_{\mathbf{x}}\right) KT = -\frac{2}{3}P\nabla \cdot \mathbf{u} = -\frac{2}{3}\nabla_{\mathbf{x}} \cdot \mathbf{q} - \frac{2}{3}\pi_{ij}\frac{\partial u_i}{\partial x_j}, \tag{3.48}$$

where use has been made of the Einstein convention of summation over repeated indices.

We have introduced a few new quantities: the **external force density**

$$\mathscr{F}^{ext}(\mathbf{x}, t) = \int d^3 v f(\mathbf{x}, \mathbf{v}; t) \mathbf{F}^{ext}(\mathbf{x}, \mathbf{v}; t); \tag{3.49}$$

the **pressure**

$$P = nKT; \tag{3.50}$$

the **anisotropic pressure tensor**

$$\pi_{ij}(\mathbf{x}, t) = m \int d^3 v [\hat{v}_i \hat{v}_j - \hat{v}^2 \delta_{ij}/3)] f(\mathbf{x}, \mathbf{v};); \quad \hat{\mathbf{v}} = \mathbf{v} - \mathbf{u}, \tag{3.51}$$

and the **heat flux**:

$$\mathbf{q}(\mathbf{x}, t) = \frac{m}{2} \int d^3 v \hat{v}^2 \hat{\mathbf{v}} f(\mathbf{v}, \mathbf{x}; t). \tag{3.52}$$

We shall see in the next sections that the pressure and the heat flux, defined in Eqs. (3.50) and (3.52), have all the dynamical properties that are to be expected from their names. We shall also see that the anisotropic pressure tensor is responsible for the viscous stresses. The anisotropic pressure tensor and the heat flux represent the so called **non-ideal** component of the fluid equations; the fluid equations obtained setting these terms equal to zero, are called the equations of **ideal hydrodynamics**.

Now, however, we have a problem. In principle, we could use Eqs. (3.46–3.48) to calculate the fields n, $\mathbf{u}$ and T, but we realize immediately that Eqs. (3.51) and (3.52) cannot be used to express π_{ij} and $\mathbf{q}$ as function of n, $\mathbf{u}$ and T, if the form of f is unknown. In fact, Eqs. (3.46–3.48) are arranged in an unclosed hierarchical structure, similar to the one for the distributions f_k, discussed in Sect. 3.2. It is easy to check that the new quantities π_{ij} and $\mathbf{q}$ are identically zero in the case of a Maxwell-Boltzmann distribution. They account in fact for the non-Maxwellian part of f that still we do not know. The only thing that we know, through Eq. (3.38), is that $g = f - f^{MB}$, and that therefore also the non-ideal terms π_{ij} and $\mathbf{q}$ are generated by the collisions.

3.9 Viscosity and Thermal Diffusivity

In order for the fluid equations (3.46–3.48) to form a closed set of equations, it is necessary to express the non-ideal terms π_{ij} and $\mathbf{q}$ as functions of n, $\mathbf{u}$ and T. This requires us to determine the non-Maxwellian contribution to the distribution $g = f - f^{MB}$. One possibility is perturbative expansion, in powers of λ/L, of the of the collision term $(\partial f / \partial t)_{coll}$ [see Eq. (3.31)]. This approach goes under the name of **Chapman-Enskog expansion** (see e.g. [K. Huang, Statistical Mechanics, 1st edn (Wiley and Sons, 1963)]).

An alternative possibility is to pursue an heuristic approach, along the line of the discussion in Sect. 3.6, to explain the origin of the non-Maxwellian component $g = f - f^{MB}$, as a superposition of Maxwellians coming from regions of the fluid separated by λ.

Let us start by estimating the heat flux in x_1, generated by a gradient of T along x_1. Of the molecules at x_1, a part will have originated from collisions at $x_1 + \lambda$, in a region at temperature $T(x_1 + \lambda)$; another at $x_1 - \lambda$, where the temperature was $T(x - \lambda)$. From the definition (3.52) for the heat flux, we obtain the estimate (recall $\hat{\mathbf{v}} = \mathbf{v} - \mathbf{u}$):

$$\mathbf{q}(x_1) = \frac{1}{2}n(x_1)m\langle \hat{v}^2\hat{\mathbf{v}} \mid x_1 \rangle \equiv \frac{m}{2} \int d^3v\, v\hat{v}^2\hat{\mathbf{v}}g(x_1, \mathbf{v})$$

$$\sim Kn[v_{th}(x_1 - \lambda)T(x_1 - \lambda) - v_{th}(x_1 + \lambda)T(x_1 + \lambda)]\mathbf{e}_1 \tag{3.53}$$

where $\pm v_{th}(x_1 \mp \lambda)$ is the component along $\mathbf{e}_1$ of the velocity of the molecules arriving at x_1 from collisions at $x_1 \mp \lambda$ (obviously, to arrive at x_1 from the left, the velocity has to be positive, and vice versa).

We can Taylor expand Eq. (3.53) in λ/L, to find $\mathbf{q}(x_1) \sim -n\lambda(\partial/\partial x_1)(v_{th}KT)$ $\mathbf{e}_1 \sim n\lambda v_{th}(\partial KT/\partial x_1)\mathbf{e}_1$. Taking the dependence of T in a generic direction:

$$\mathbf{q} = -n\kappa_T \nabla KT, \qquad \kappa_T \sim \lambda v_{th} \sim \frac{\lambda^2}{\tau_{coll}}, \tag{3.54}$$

where the coefficient κ_T is called the **thermal diffusivity** for the gas. The flux-relation described by Eq. (3.54) is in fact that of a diffusion process, analogous to the one discussed in Sect. 2.7.1.

We proceed in similar way with the anisotropic pressure tensor, starting from the observation that, in this case, in place of a thermal energy flux, we have a flux of linear momentum. As the heat flux was generated by a gradient of temperature, the momentum flux is generated by a gradient of velocity $\partial u_i/\partial x_j$. This velocity gradient has components both of shear type $i \neq j$, and of local deformation/compression type $i = j$.

In the case of shear, what we have is basically a sliding of fluid masses with respect to one another. Let us imagine that passing from $x_1 - \lambda$ to $x_1 + \lambda$ there is a variation in the component u_2 of the fluid velocity.

In this case, the molecules that arrive at x_1, will have a component of momentum $mu_2(x_1 \pm \lambda)$. We find

$$\pi_{12}(x_1) = nm\langle v_1 v_2 \mid x_1 \rangle \equiv m \int d^3v\, v v_1 v_2 g(x_1, \mathbf{v})$$

$$\sim nm[v_{th}(x_1 - \lambda)u_2(x_1 - \lambda) - v_{th}(x_1 + \lambda)u_2(x_1 + \lambda)].$$

where again $\pm v_{th}(x_1 \mp \lambda)$ is the component along $\mathbf{e}_1$ of the velocity of the molecules arriving at x_1 from $x_1 \mp \lambda$.

Taylor expanding in λ/L, we find $\pi_{12}(x_1) \sim -n\lambda(\partial/\partial x_1)(v_{th}u_2) \sim n\lambda v_{th}$ $(\partial u_2/\partial x_1)$. We know however, as it is clear from Eq. (3.51), that the anisotropic pressure tensor is symmetric and has zero trace. To make this condition satisfied, we must therefore add to $\pi_{ij} \sim \partial u_j/\partial x_i$ the necessary velocity derivatives components.

The corrected expression is

$$\pi_{ij} = -nmv\left(\frac{\partial u_i}{\partial x_j} + \frac{\partial u_j}{\partial x_i} - \frac{2}{3}\nabla \cdot \mathbf{u}\right); \quad \mu = nmv; \quad v \sim \lambda v_{th} \sim \frac{\lambda^2}{\tau_{coll}}. \quad (3.55)$$

The coefficients v and μ are called, respectively, the **kinematic viscosity** and the **dynamic viscosity** of the gas. As the heat flux is associated with diffusion of thermal energy, the anisotropic pressure tensor is associated with diffusion of linear momentum, with the viscosity v playing the role of momentum diffusivity. The two coefficients κ_T and v have in fact, in many situations, the same magnitude. In the case of air, at atmospheric conditions of temperature and pressure, we have $\kappa_T \sim v \sim 0.1 \text{cm}^2/\text{s}$.

Substituting equations, such as (3.54) and (3.55), into the moment Eqs. (3.46–3.48), we obtain the equations of viscous hydrodynamics. From Eq. (3.47), we obtain the **Navier-Stokes** equation

$$nm\left(\frac{\partial}{\partial t} + \mathbf{u} \cdot \nabla\right)\mathbf{u} + \nabla P = \mathscr{F}^{ext} + \mu\left(\nabla^2\mathbf{u} + \frac{1}{3}\nabla\nabla \cdot \mathbf{u}\right) \quad (3.56)$$

From Eq. (3.48), we obtain the **heat transport** equation

$$\frac{3}{2}n\left(\frac{\partial}{\partial t} + \mathbf{u} \cdot \nabla\right)KT + P\nabla \cdot \mathbf{u} = \frac{2}{3}n\kappa_T\nabla^2 KT - \pi_{ij}\frac{\partial u_i}{\partial x_j}. \quad (3.57)$$

3.9.1 Diffusion in a Gas Mixture

Not only do heat and momentum diffuse in a gas. Also chemical substances undergo this process, that is the adaptation, to the case of a gas, of the classical problem of the drop of ink that dissolves in a quiescent liquid.

Let us consider the mixture of two non-reactive chemical species, a and b, of which the first one is supposed very dilute: $\varrho_a \ll \varrho_b \simeq \varrho$; here, $\varrho_{a,b} = n_{a,b}m_{a,b}$ are the mass densities of the two substances and $\varrho = \varrho_a + \varrho_b$. We can define a fluid velocity $\mathbf{u}$, starting from the linear momentum density $\varrho_a\mathbf{u}_a + \varrho_b\mathbf{u}_b = \varrho\mathbf{u}$, and we find again $\mathbf{u} \simeq \mathbf{u}_b$. Let us consider a thermodynamic equilibrium condition for all the macroscopic variables of the system, except for the concentration n_a, that is supposed to be a function of position. Since at equilibrium $\mathbf{u}$ is spatially uniform, we can choose to work in the reference frame in which $\mathbf{u} = 0$. The velocity $\mathbf{u}_a$ becomes therefore the dispersion velocity of substance a in the mixture; knowing $\mathbf{u}_a$ would allow us to determine the dynamics of the mixing process through evolution for the concentration n_a, which, in the absence of chemical reactions, is governed by the continuity equation

$$\partial_t n_a + \nabla \cdot (n_a\mathbf{u}_a) = 0. \quad (3.58)$$

The velocity $\mathbf{u}_a$ is the result of the collisions of molecules of species a with the molecules of species b. We can estimate $\mathbf{u}_a$ from considerations analogous to the those utilized in the previous section, to obtain expressions for the thermal diffusivity of a single species.

Suppose that n_a has a gradient along x_1. The flux of species a molecules in x_1, that previously collided with species b molecules at $x_1 \pm \lambda_a$, will be:

$$n_a(x_1, t))u_a(x_1, t) \sim n_a(x_1 - \lambda_a, t)v_{th,a} - n_a(x_1 + \lambda_a, t)v_{th,a}$$
$$\simeq -\lambda_a v_{th,a} \frac{\partial n_a(x_1)}{\partial x_1}, \qquad (3.59)$$

where λ_a is the mean free path for species a. We find therefore a flux-gradient relation in the form $n_a u_a = -\kappa_a \nabla n_a$, where $\kappa_a \sim \lambda_a v_{th,a}$ is the diffusivity of species a in the mixture. Substituting into Eq. (3.58), we find the diffusion equation:

$$\partial_t n_a = \kappa_a \nabla^2 n_a. \qquad (3.60)$$

The analogy with the behavior of the random walker discussed in Sect. 2.7.1 is even more apparent than in the case of heat diffusion, as in the present case, we have a real random walk of species a molecules, induced by the collisions with the other molecules in the gas.

The presence of a mean relative velocity between species, must be associated with some force acting between the two. Let us determine its nature.

The Navier-Stokes equation for species a, will have the same expression (3.56), as for the case of a single species, except for a new force density term produced by collisions with molecules b. This force density can be estimated as $\mathscr{F}^{coll} \sim -\varrho_a u_a / \tau_{coll}$. Consider the simpler situation of a density profile, that, at time t, is linear: $n_a(\mathbf{x}, t) \simeq n_{0,a} + \mathbf{a} \cdot \mathbf{x}$, so that u_a is spatially homogeneous. This leads to the relation[4]

$$\nabla P_a + \varrho_a u_a / \tau_{coll} = 0,$$

where $P_a = n_a T$ is called the **partial pressure** of species a. Solving with respect to u_a, we find again Eq. (3.59). We see therefore that the dispersion velocity u_a is nothing but the reaction of species a to the partial pressure gradient produced by the spatial inhomogeneity in the profile of n_a.

3.10 Elementary Transport Theory

The fluid equations (3.46–3.48) can be seen as conservation equations for the mass, momentum and energy, in the fluid parcels, as they are transported by the flow field $\mathbf{u}(\mathbf{x}, t)$. To illustrate the idea, let us introduce the concept of **fluid element**, that is a

---

[4] Absence of the term $\varrho_a \partial_t$ in the equation, is due to Eq. (3.59) and to the fact that n_a is linear; from (3.60): $\partial_t n_a = 0$.

volume V_e of fluid, bounded by a surface S whose points move with the fluid velocity $\mathbf{u}(\mathbf{x}, t)$.

By construction, the mass M_e contained in V_e, does not change. Thus, mass conservation is trivial.

Let us pass to consider momentum conservation. Assume that the volume element V_e is infinitesimal, so that $\mathbf{u}$ and T are approximately constant inside. The fact that the mass M_e in V_e is constant, allows us to write

$$M_e \dot{\mathbf{u}}(t) = M_e \lim_{\Delta t \to 0} \frac{1}{\Delta t}[\mathbf{u}(\mathbf{x} + \mathbf{u}(\mathbf{x}, t)\Delta t, t + \Delta t) - \mathbf{u}(\mathbf{x}, t)]$$

$$= M_e \left(\frac{\partial}{\partial t} + \mathbf{u}(\mathbf{x}, t) \cdot \nabla\right)\mathbf{u}(\mathbf{x}, t). \tag{3.61}$$

Here, $\mathbf{u}(t) \equiv \mathbf{u}(\mathbf{x}(t), t)$ is the fluid velocity measured at the fluid element position $\mathbf{x}(t)$, with $\dot{\mathbf{x}}(t) = \mathbf{u}(t) \equiv \mathbf{u}(\mathbf{x}(t), t)$. Thus, the momentum equation (3.47) is in the form of a density for the second law of Newton: $M_e \dot{\mathbf{u}} = F_e$, where F_e is the total force exerted on the volume element.

We can repeat the calculation leading to Eq. (3.61) with the thermal energy in the fluid element

$$E_e^{th} = (3/2)nV_eKT. \tag{3.62}$$

We obtain the result

$$\dot{E}_e^{th}(t) = \frac{3}{2}nV_eK\left(\frac{\partial}{\partial t} + \mathbf{u}(\mathbf{x}, t) \cdot \nabla\right)T(\mathbf{x}, t). \tag{3.63}$$

Thus, Eq. (3.57) is in the form of a heat budget equation for the volume element.

The operator

$$\frac{D}{Dt} \equiv \frac{\partial}{\partial t} + \mathbf{u}(\mathbf{x}, t) \cdot \nabla \tag{3.64}$$

appearing in both Eqs. (3.61) and (3.57), is called **material derivative** and is basically a time derivative measured in the reference frame traveling with the fluid. We have already seen that the acceleration of the fluid element at $\mathbf{x}$, can be expressed as the material derivative of the fluid velocity at that point: $\dot{\mathbf{u}}(t) = D\mathbf{u}(\mathbf{x}, t)/Dt$. We can use the material derivative D/Dt to rewrite the continuity equation (3.46) in the form

$$\frac{Dn(\mathbf{x}, t)}{Dt} = -n(\mathbf{x}, t)\nabla \cdot \mathbf{u}(\mathbf{x}, t).$$

This equation is telling us that, if the flow is divergenceless, the density will remain constant as it is transported by the flow. This is the kind of situation that one would observe if, say, droplets of oil were transported by a water flow: were there is water, the density will remain that of water, were there is oil, the density will remain that of oil. The fluid elements neither contract nor expand as they are advected by the flow.

This tells us that imposing the zero divergence condition

$$\nabla \cdot \mathbf{u}(\mathbf{x}, t) = 0, \tag{3.65}$$

is equivalent to impose incompressibility of the flow.

3.10.1 Balance of Forces

Let us study the balance of forces on a fluid element V_e. Let us consider V_e small enough, that the variation of P $\mathbf{u}$ and $\mathscr{F}^{ext}$ inside, are themselves small. Substituting Eq. (3.61) into Eq. (3.56), we find (we consider for simplicity the case of an incompressible flow $\nabla \cdot \mathbf{u} = 0$):

$$M_e \dot{\mathbf{u}} = \mathbf{F}_e = \int_{V_e} d^3x \, [-\nabla P + \mu \nabla^2 \mathbf{u} + \mathscr{F}^{ext}]. \tag{3.66}$$

We verify at once that the field P in Eq. (3.66) is indeed the fluid pressure, as the definition, Eq. (3.50), suggests. Taking a volume element in the shape of a cube, whose sides are Δx_i, $i = 1, 2, 3$, and focusing on the component 1 of the force, we find in fact:

$$-\int_{V_e} d^3x \, \frac{\partial P}{\partial x_1} \simeq -\Delta x_2 \Delta x_3 [P(x_1 + \Delta x_1) - P(x_1)],$$

that is just the difference of the pressure forces on the sides $\Delta x_1 \Delta x_3$ of the cube, at x_1 and $x_1 + \Delta x_1$.

The identification of P as a real fluid pressure, and of T as a real thermodynamic temperature, allows to interpret Eq. (3.50) as the **equation of state** of the ideal gas.

Let us pass to analyze the viscous force. Using the divergence theorem, we can write

$$\mu \int_{V_e} d^3x \, \nabla^2 \mathbf{u} = \mu \int_S dS_i \frac{\partial \mathbf{u}}{\partial x_i}.$$

If we focus on a situation in which $\mathbf{u} = (0, u_2(x_1), 0)$, the only component of the force, that will survive, is the one along x_2:

$$\Delta x_2 \Delta x_3 \, \mu \left[\frac{\partial u_2(x_1 + \Delta x_1)}{\partial x_i} - \frac{\partial u_2(x_1 + \Delta x_1)}{\partial x_i} \right],$$

that is the difference between the two tangential forces at the sides $\Delta x_1 \Delta x_3$ of the cube, at x_1 and $x_1 + \Delta x_1$. Thus, a velocity component u_2 that becomes larger at large x_1 will lead to a push upwards on V_e at the side at $x_1 + \Delta x_1$, and downwards at the side at x_1, that is precisely the behavior expected from a viscous force.

3.10.2 Thermal Energy Budget

Proceeding as in the previous section, we can derive a local balance equation for the thermal energy. Substituting Eq. (3.63) into Eq. (3.57), we obtain

$$\dot{E}_e^{th} = \int_{V_e} d^3x \left[-P\nabla \cdot \mathbf{u} - \pi_{ij}\frac{\partial u_i}{\partial x_j} + \kappa_T \nabla^2 T \right].$$

We can use the divergence theorem, to transform the term involving $\nabla \cdot \mathbf{u}$ into a surface integral. Taking V_e small, so that P is almost constant in its interior:

$$\int_{V_e} d^3x\, P\nabla \cdot \mathbf{u} \simeq P \int_{S_e} d\mathbf{S} \cdot \mathbf{u} = P\dot{V}_e.$$

Using again the diverge theorem to convert the heat diffusion term into a surface integral, we obtain

$$\dot{E}_e^{th} = -P\dot{V}_e - V_e\pi_{ij}\frac{\partial u_i}{\partial x_j} + \kappa_T \int_{S_e} d\mathbf{S} \cdot \nabla T. \tag{3.67}$$

We recognize in Eq. (3.67) a version of the first law of thermodynamics, in which the first contribution is the heating/cooling of the fluid element in compression/expansion processes; the second contribution is the heat produced by the viscous forces, that is positive defined, as required for a dissipative process; the third is the heat received through diffusion from the outside, that will be positive if the temperature outside is higher, negative if it is lower.

3.10.3 Relaxation to Thermal Equilibrium

The fluid equations (3.46), (3.56) and (3.57) can be utilized to describe the relaxation to thermodynamic equilibrium of a fluid, in the absence of external driving.

If there are initially pressure unbalance in the fluid, the relaxation process will have a mechanical component, that could be estimated from dimensional analysis. Assuming a single space scale L for the pressure gradient, the kinetic, pressure and viscosity terms in Eq. (3.66) will have order of magnitude

$$\dot{\mathbf{u}} \sim \frac{L}{\tau^2}, \quad \frac{1}{M}\int_{V_e} d^3x\, \nabla P \sim \frac{\Delta P}{L\varrho}, \quad \frac{\mu}{M}\int_{V_e} d^3x\, \nabla^2\mathbf{u} \sim \frac{\nu}{L\tau}, \tag{3.68}$$

where τ is the relaxation time of the process. The mechanical component in the relaxation process will be important when the kinetic term in Eq. (3.68) is much larger than the viscous one. This leads to the estimate of the relaxation time

$$\tau_{mech} \sim L\sqrt{\frac{\varrho}{\Delta P}}, \tag{3.69}$$

to be compared with the viscous time scale

$$\tau_{visc} \sim \frac{L^2}{\nu}, \tag{3.70}$$

that would characterize a dynamics in which the fluid motion is negligible. The relative importance of the mechanical and diffusive contribution, is given by the (inverse) ratio of the respective time scale, typically referred to as the Reynolds number

$$Re = \frac{\tau_{visc}}{\tau_{mech}} \sim \frac{L}{\nu}\sqrt{\frac{\Delta P}{\varrho}}. \tag{3.71}$$

Fluid motions will become more pronounced if the pressure gap and the scale of the inhomogeneity in the fluid are large.

Now, if viscosity were zero, there would be a continuous conversion of pressure gaps, into fluid motions, and then again into other pressure gaps, without an end. In general, we may have also temperature inhomogeneities in fluid, and, if initially there were none, they would be generated by local compressions and expansions in the fluid. Again, if the thermal diffusivity were zero, no heat could be transferred between fluid elements, and the temperature inhomogeneities would form and destroy continuously in the fluid. Hence, an ideal fluid could never reach thermal equilibrium.

This would suggest that the ideal processes taking place during relaxation, accompany, but are not essential for thermodynamic equilibrium. This of course cannot be the case. In fact, the fluid motions generated in a non-equilibrium condition, not only push fluid parcels around, but produce **mixing** of quantities such as the temperature. By mixing, we mean the process whereby fluid regions with different characteristics, say temperature, become more and more intertwined, in such a way that, in the end, in any region there will be both hot and cold spots present. The process is illustrated in Fig. 3.4. We stress the difference with the process of homogenization by diffusion, in which the temperature in any given region becomes really uniform (pushing the analogy to the limit, we could see diffusion as a sort of mixing at molecular scales).

In fact, during mixing, entropy does not change; this is particularly simple to see, using Eq. (3.39), in the case the area of the hot and cold regions in the mixed and unmixed state remain equal (as it would occur if mixing were produced by an incompressible flow).

Mixing, however, causes the characteristic scale of the filaments (the typical separation between cold and hot regions) to becomes rapidly very small. At this point, thermal diffusion will start to transfer heat among them, much more efficiently than if these spots had been widely separated. This has the important consequence that the

Fig. 3.4 Mixing of a dark syrup in clear water. Notice the increasingly thinner filaments produced in the process (Image taken from physicsworld.com; courtesy of J.-L. Thiffeault)

homogenization time scale is not that in Eq. (3.70), rather, it is the time required for the filaments to become very small. This squeezing time, in most situations, could be estimated as the mechanical time scale of Eq. (3.69).

3.11 Problems

Problem 1 The microorganisms in a Petri dish have individual death rate Γ_D $(A \to \varnothing)$ and reproduction rate by cell division $(A \to AA)$ Γ_B. Indicate by $N(t)$ the number of individuals in the dish at time t, and assume the birth and death rates $\Gamma_{B,D}$ to be constant and independent of N.

- Write the master equation for the PDF $\rho(N, t)$.
- Derive the evolution equation for the mean $\langle N(t) \rangle$, and for the second moment $\langle N^2(t) \rangle$.
- Discuss the neutral case $\Gamma_B = \Gamma_D$.

Problem 2 Consider again the microorganism colony of Problem 1. We would like to include some limitation mechanism for population growth. We can model this effect by an increase of the death rate at large N. The simplest possible expression is $\Gamma_D = \Gamma_B N/N_0$.

- Write again the master equation for the PDF $\rho(N, t)$.
- Derive again the evolution equation for mean $\langle N(t) \rangle$, and for the second moment $\langle N^2(t) \rangle$ and determine their values at stationary state.
- In which range of values of the parameter N_0, could we expect a mean-field approximation to be applicable? Work perturbatively in the fluctuation $N - \langle N \rangle$.

Problem 3 Show that substituting the local Maxwell-Boltzmann distribution (3.35) into the Boltzmann (3.30), leads to the equations of ideal hydrodynamics.

Problem 4 How should the Boltzmann equation be modified to describe a gas of two chemical species A and B in which molecules A convert with individual reaction rate Γ into molecules B? Write the continuity equations for the two species A and B.

Problem 5 The space shuttle was approximately 40 m long. Estimate the height in the atmosphere at which a fluid description of its aerodynamic properties, ceases to be possible (for simplicity, suppose that both atmospheric temperature and chemical composition remains constant with respect to height). How does the result depend on the speed of the vehicle?

Problem 6 Calculate the force necessary to make a flat body of area A slide at velocity $\bar{u}$ parallel to a plane surface. Consider the gap h between the body and the plane to be very small compared to the horizontal size of the body. Disregard the forces necessary to keep the body at separation h from the plane. Assume laminar flow conditions in the gap.

Solution Consider the volume V in the gap between flat body and surface, as indicated in Fig. 3.5. The horizontal force F_A from the flat body (on top) on the fluid, is the force that must be exerted from the outside on the flat body itself to keep it moving. The total force on V is obtained integrating the Navier-Stokes equation

$$F_V = \int_V d^3x \left[-\varrho u \frac{\partial u}{\partial x} - \frac{\partial P}{\partial x} + \mu \nabla^2 u \right].$$

We do not apply pressure laterally to V, so that P is independent of x. Similarly for u, so that

$$F_V = \mu \int_V d^3x \, \nabla \cdot (\nabla u) = \mu \int_{\partial V} d\mathbf{S} \cdot \nabla u.$$

We are interested in the contribution to F_V from the side in contact with the upper slab

$$F_A = \mu A \frac{\partial u}{\partial y}\bigg|_{y=h}$$

(the total force on V must be zero, as the fluid in it does not accelerate). To calculate the force, we must know the velocity profile $u(y)$. This can be obtained from the Navier Stokes equation, imposing stationarity of the flow and absence of horizontal

Fig. 3.5 Flow geometry for Problem 5

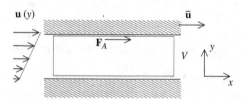

pressure gradients. We obtain simply

$$\frac{\partial^2 u}{\partial y^2} = 0 \Rightarrow u(y) = \frac{\bar{u}y}{h},$$

that gives for the force

$$F_A = \frac{\mu A \bar{u}}{h}.$$

3.12 Further Reading

Simple derivations of the BBGKY hierarchy, kinetic equations and fluid equations:

- S. Ichimaru, *Basic Principles of Plasma Physics: A Statistical Approach* (Benjamin, 1973)
- K. Huang, *Statistical Mechanics*. 2nd edn (Wiley and Sons, 1987)

Reference books for kinetic theory in general:

- R.L. Liboff, *Kinetic Theory*. 2nd edn (Wiley and Sons, 1998)
- C. Cercignani, *The Boltzmann Equation and Its Applications* (Springer, 1988)
- L.D. Landau, E.M. Lifsits, *Physical Kinetics* (Pergamon Press, 1981)

For fluid mechanics and related arguments, see e.g.:

- P. Kundu, *Fluid Mechanics* (Academic Press, 1990)

Appendix

A.1 Derivation of the Moment Equations

All the three conservation equations (3.46), (3.47) and (3.48) can be derived taking moments of the Boltzmann equation (3.30), as indicated in Eq. (3.45). The fact that the contribution from $(\partial f/\partial t)_{coll}$ drops off the continuity equation (3.46), is not casual, and in fact is a consequence of conservation of the number of molecules in collisions. In the same way, it is possible to prove that conservation of momentum and energy in collisions makes the contribution from $(\partial f/\partial t)_{coll}$ equal to zero also in Eqs. (3.47) and (3.48). Writing in compact form:

$$\int d^3 v I_k(\mathbf{v}) \left(\frac{\partial f}{\partial t}\right)_{coll} = 0, \quad k = 0, 1, 2, \tag{3.72}$$

where $I_0 = 1$, $I_1 = m\mathbf{v}$ and $I_2 = \frac{mv^2}{2}$.

We prove Eq. (3.72) proceeding as in Sect. 3.5.3. Conservation of the three quantities I_k, $k = 0, 1, 2$, in a collision, can be expressed as

$$I_k(1) + I_k(2) = I_k(1') + I_k(2'), \tag{3.73}$$

where, in the notation of Sect. 3.5.3 and following, $I_k(1) \equiv I_k(\mathbf{v}_1)$, etc. Using Eq. (3.31), we have trivially:

$$\int d^3 v_1 \, I_k(\mathbf{v}_1) \left(\frac{\partial f(\mathbf{x}, \mathbf{v}_1; t)}{\partial t} \right)_{coll}$$

$$= \int d\mu_{\mathbf{v}} w(\mathbf{v}'_{1,2} \to \mathbf{v}_{1,2}) [f(1')f(2') - f(1)f(2)] I_k(1)$$

$$= \int d\mu_{\mathbf{v}} w(\mathbf{v}'_{1,2} \to \mathbf{v}_{1,2}) [f(1')f(2') - f(1)f(2)] I_k(2)$$

$$= \int d\mu_{\mathbf{v}} w(\mathbf{v}_{1,2} \to \mathbf{v}'_{1,2}) [f(1)f(2) - f(1')f(2')] I_k(1')$$

$$= \int d\mu_{\mathbf{v}} w(\mathbf{v}_{1,2} \to \mathbf{v}'_{1,2}) [f(1)f(2) - f(1')f(2')] I_k(2'). \tag{3.74}$$

Exploiting the symmetry of the collision kernel, $w(\mathbf{v}'_{1,2} \to \mathbf{v}_{1,2}) = w(\mathbf{v}_{1,2} \to \mathbf{v}'_{1,2})$, we obtain therefore, summing the four identical contribution in the lines of Eq. (3.74), and exploiting Eq. (3.73):

$$\int d^3 v_1 \, I_k(1) \left(\frac{\partial f}{\partial t} \right)_{coll} = \frac{1}{4} \int d^3 v_1 \, w(\mathbf{v}'_{1,2} \to \mathbf{v}_{1,2})$$

$$\times [f(1')f(2') - f(1)f(2)][I_k(1) + I_k(2) - I_k(1') - I_k(2')] = 0.$$

At this point, we can proceed with analysis of the LHS of Eq. (3.45), for $k = 1$ and $k = 2$:

$$\int d^3 v I_k(\mathbf{v}) \left[\frac{\partial f}{\partial t} + \mathbf{v} \cdot \nabla_x f + \frac{1}{m} \nabla_{\mathbf{v}} \cdot (\mathbf{F}^{ext} f) \right] = 0. \tag{3.75}$$

A.1.1 Momentum Equation

This corresponds to the case $k = 1$. The three terms in Eq. (3.75), read, respectively:

$$\int d^3 v m \mathbf{v} \frac{\partial f}{\partial t} = \frac{\partial nm\mathbf{u}}{\partial t}, \tag{3.76}$$

$$\int d^3 v m v_i v_j \frac{\partial f}{\partial x_j} = \frac{\partial}{\partial x_j} \left\{ nm \left[u_i u_j + \langle (v_i - u_i)(v_j - u_j) \rangle \right] \right\}, \tag{3.77}$$

$$\int d^3 v v_i \frac{\partial F_j^{ext} f}{\partial v_j} = -\int d^3 v F_i f = \mathscr{F}^{ext}, \tag{3.78}$$

where use has been made of Eqs. (3.44) and (3.49). We can combine Eq. (3.76) and the first term to RHS of Eq. (3.77) to produce

$$\frac{\partial nmu_i}{\partial t} + \frac{\partial nmu_i u_j}{\partial x_j} = nm\left(\frac{\partial}{\partial t} + \mathbf{u} \cdot \nabla\right) u_i, \tag{3.79}$$

where use has been made of the continuity equation (3.46). The average appearing in the second term to RHS of Eq. (3.77) can be rewritten:

$$nm\langle(v_i - u_i)(v_j - u_j)\rangle = nm\Big[\langle|\mathbf{v} - \mathbf{u}|^2\rangle\delta_{ij}/3$$
$$+ \langle[(v_i - u_i)(v_j - u_j) - |\mathbf{v} - \mathbf{u}|^2\delta_{ij}/3]\rangle\Big] = P\delta_{ij} + \pi_{ij}, \tag{3.80}$$

where use has been made of Eqs. (3.44), (3.49), (3.50) and (3.51). Substituting Eqs. (3.79), (3.80) and (3.78), into Eq. (3.75), for $k = 1$, we obtain the momentum equation (3.47).

A.1.2 Energy Equation

We pass to consider the $k = 2$ case. The three terms in Eq. (3.75) will read now:

$$\int d^3 v \frac{mv^2}{2} \frac{\partial f}{\partial t} = \frac{\partial}{\partial t}\left[\frac{nmu^2}{2} + \frac{3nKT}{2}\right], \tag{3.81}$$

$$\int d^3 v \frac{mv^2}{2} v_j \frac{\partial f}{\partial x_j} = \frac{\partial}{\partial x_i}\left\{\frac{nmu_i u^2}{2} + \frac{3}{2}u_i nKT + u_j\left[P\delta_{ij} + \pi_{ij}\right] + q_i\right\}, \tag{3.82}$$

$$\int d^3 v \frac{v^2}{2}\nabla_{\mathbf{v}} \cdot (\mathbf{F}^{ext} f) = -\left[\mathbf{u} \cdot \mathscr{F}^{ext} + n\langle\hat{\mathbf{v}} \cdot \mathbf{F}^{ext}\rangle\right], \tag{3.83}$$

where $\hat{\mathbf{v}} = \mathbf{v} - \mathbf{u}$, and use has been made of Eqs. (3.44) and (3.49–3.52).

It is possible to see that, if the dependence of $\mathbf{F}^{ext}$ on $\mathbf{v}$, is that of a Lorentz force, or if F^{ext} is independent of velocity, then $\langle\hat{\mathbf{v}} \cdot \mathbf{F}^{ext}\rangle = 0$.

We can then subtract, from Eqs. (3.81–3.83), the corresponding terms, obtained by scalar product of the momentum Eq. (3.47) and $\mathbf{u}$. This corresponds to subtracting the mechanical component from the total energy balance equation, obtained setting $k = 2$ in Eq. (3.75).

The advective terms in Eq. (3.47) give

$$\frac{nm}{2}\left(\frac{\partial}{\partial t} + \mathbf{u} \cdot \nabla\right)u^2 = \frac{m}{2}\left(\frac{\partial nu^2}{\partial t} + \nabla \cdot (n\mathbf{u}u^2)\right),$$

that cancel the u^2 terms in Eqs. (3.81) and (3.82). The pressure term in Eq. (3.47), produces a term

$$u_i \frac{\partial}{\partial x_i}\left(P\delta_{ij} + \pi_{ij}\right),$$

that, subtracted to Eq. (3.82) leaves

$$\left(P\delta_{ij} + \pi_{ij}\right)\frac{\partial u_j}{\partial x_i}.$$

The work term $\mathbf{u} \cdot \mathbf{f}^{ext}$, in Eq. (3.83), is cancelled by the corresponding term arising from the momentum equation. Grouping the surviving terms in Eqs. (3.81) and (3.82), gives finally the energy Eq. (3.48).

Chapter 4
Thermodynamics

4.1 Basic Definitions

It can be said that the goal of thermodynamics is to describe the basic properties
and the dynamics of physical systems, in terms of a reduced set of "macroscopic"
variables. These variables can be grouped into two classes:

- **Intensive variables**, such as pressure and temperature, that describe local properties of the system, and become independent of position at equilibrium (in the
  absence of external forces).
- **Extensive variables**, such as volume and entropy, that characterize the system
  as a whole, and have additive character; for instance, the total volume of two
  distinct systems A and B, that is obviously the sum of the two volumes $V_{A,B}$:
  $V_{A \cup B} = V_A + V_B$.

An intensive quantity, due to its local character, could always be expressed as a
function only of other intensive parameters. For instance, for the pressure of an
ideal gas: $P = nT$, where both the density n and the temperature T are intensive.
Conversely, an extensive quantity could always be expressed as a product of an
extensive and an intensive quantity, where the second acts as a density with respect
to first. For instance: $N = nV$.

We have seen that, to be defined in a meaningful way, all these variables require
that a condition of thermodynamic equilibrium be satisfied at least locally. The picture of a macroscopic system that is provided in this way is that of a collection of
subsystems internally in equilibrium, even though they may be not in equilibrium
among themselves.

It is clear that considering a macroscopic object (or part of an object), that interacts with other macroscopic systems in a time-dependent situation, as something
internally in equilibrium, may be problematic. A detailed description of the internal
interactions that govern the system dynamics is therefore required. This is the level
of description provided e.g. by the equations of fluid dynamics. The alternative is to
disregard all system internal details and to describe its dynamics as a sequence of

© Springer International Publishing Switzerland 2015
P. Olla, *An Introduction to Thermodynamics and Statistical Physics*,
UNITEXT for Physics, DOI 10.1007/978-3-319-06188-7_4

transitions between equilibrium states. Such a level of description may have sense only if:

- The system evolution is very slow, so that its condition at every instant can be approximated as thermal equilibrium.
- The system is not in equilibrium at all, except at the initial and final state, but we are not interested in its dynamics, rather, in the final outcome of the processes taking place.

Processes of the first kind are called **quasistatic**. An example could be the slow isothermal compression of an ideal gas, such that the gas pressure P is determined instantaneously by the volume V as $P(t) = NKT/V(t)$. Processes that are not quasistatic, are characterized by the fact that the system ceases to be spatially homogeneous, which obviously implies that an instantaneous description in terms of global thermodynamic variables is not possible.

The **free expansion process**, already discussed in Sect. 3.3, is an example of such a process, and of the way in which purely thermodynamic considerations allow to determine its outcome, without any regard to the dynamics.

Imagine that a thermally isolated volume V is subdivided into two parts, separated by a wall. One of the parts contains a gas at given temperature and pressure, while the other is empty. Suppose that the wall is instantaneously eliminated, so that the gas is left free to expand. The expansion process is likely to be violent, with turbulence and possible appearance of shock waves; obviously, it could not be described in terms of just a few thermodynamic variables. We can nevertheless determine the final (equilibrium) state, from the properties of the initial (equilibrium) state. Conservation of energy gives us in fact $T_{final} = T_{initial} \equiv T$, and therefore $P_{final} = PV_A/V$.

Another aspect that characterizes a thermodynamic transformation, is its degree of **irreversibility**, i.e. how much the total entropy of the system, and of that part of the outside world in interaction with it, will rise during the process. Strictly speaking, any process in an isolated system will be irreversible, that is just a manifestation of the tendency of the system to approach thermodynamic equilibrium. The same concept applies to an arrangement of interacting systems, that constitute together an isolated macrosystem. The simplest example, already considered in Sect. 3.7, is that of two volumes of gas, A and B, that exchange heat at constant volume, and are thermally insulated from the outside world. An important case is that in which one of the two subsystems is much larger than the other, so that its heat capacity can be considered infinite: a so-called **heat bath**. More complex arrangements can be envisioned, with more than one heat bath, volume changes (as in thermal engines), or even chemical reactions, phase changes, etc., In all these cases, the transformations occurring in the individual systems correspond to an entropy transfer that only in part results into a global rise of entropy.

- This suggests us that we can introduce the concept of **reversible** process, as the limit of a transformation in which the global rise of entropy is negligible, compared to the entropy exchanges taking place between the different systems.

To make the argument more quantitative, let us go back to the example in Sect. 3.7: a system composed of two parts A and B in internal equilibrium, that exchange heat at constant volume. Thus, $\Delta Q_{A,B} \equiv \Delta E_{A,B}$. We have seen that the total entropy $S = S_A + S_B$, expressed as function of the energy of one of the parts (say E_A) has a maximum at equilibrium, $E_A = \bar{E}_A$. Thus, for small deviations, $\Delta E_A = E_A - \bar{E}_A$, the total entropy S will differ from its equilibrium value $S(\bar{E}_A)$ by quadratic terms:

$$\Delta S = S(E_A) - S(\bar{E}_A) = \frac{1}{2} S''(\bar{E}_A)(E_A - \bar{E}_A)^2 + O((E_A - \bar{E}_A)^3). \qquad (4.1)$$

The variation in the entropy in the two sub-systems, however, is linear in the energy; within quadratic terms:

$$\Delta S_A = -\Delta S_B = \Delta E / T, \qquad (4.2)$$

where $\Delta S_A = S_A(E_A) - S_A(\bar{E}_A)$ and $\Delta S_B = S_B(E - E_A) - S_B(E - \bar{E}_A)$. Imposing explicitly the condition $\Delta S_{A,B} \gg \Delta S$, exploiting Eqs. (4.1) and (4.2), and using $S'_A(\bar{E}_A) = S'_B(\bar{E}_B) = 1/T$, together with $S''(\bar{E}_A) \sim -T^{-2}(\partial E_A/\partial T)^{-1}$, we find the condition

$$T \frac{\partial E_A}{\partial T} \gg \Delta E_A, \qquad (4.3)$$

that, for an ideal gas, will read simply $\Delta T_A < (T_B - T_A) \ll T_A$: in order for the transformation to be considered reversible, it is necessary that the temperature gap between the two bodies in contact, is small. The range of reversible transformations coincides therefore with the interval of values of $E_{A,B}$, for which the linear approximation is valid: $S_{A,B}(E_{A,B}) \simeq S_{A,B}(\bar{E}_{A,B}) + \Delta E_{A,B}/T$. Notice that the reversibility condition $\Delta T \ll T$ coincides in practice with a quasi-static condition for the relaxation in the systems.

A condition, analogous to Eq. (4.3), is valid for transformations involving changes of volume. In this case, it is required that the pressure difference between the two bodies be small, compared with the pressure in each of them. Such a system in internal equilibrium, that quasistatically receives heat $dQ = T dS$ from, and executes work $dW = P dV$ on the external world, will satisfy a **first law of thermodynamics** in the form

$$dE = T dS - P dV. \qquad (4.4)$$

Notice that by writing this expression, we are assuming implicitly that our independent variables are now S and V: $E = E(S, V)$. Thus

$$T = \frac{\partial E}{\partial S} \quad \text{and} \quad P = -\frac{\partial E}{\partial V}. \qquad (4.5)$$

From the first law of thermodynamics, Eq. (4.4), we can introduce important quantities such as the heat capacity at constant volume and at constant pressure:

$$C_V = \left(\frac{\dvec{Q}}{dT}\right)_V = \left(\frac{\partial E}{\partial T}\right)_V; \quad C_P = \left(\frac{\dvec{Q}}{dT}\right)_P = \left(\frac{\partial E}{\partial T}\right)_P + P\left(\frac{\partial V}{\partial T}\right)_P. \quad (4.6)$$

Notice the need to specify the quantity kept constant in the differentiations, meaning that the independent variables are taken to be, in the two cases, T, V and T, P. (In expressions such as $T = \partial E/\partial S$, the subscript V would be redundant, as the partial derivative of $E(S, V)$ with respect to S, assumes implicitly that V is kept constant). We provide the expression for the heat capacity for an ideal gas, for which $PV = NT$ and $E = (\alpha/2)NT$, with α the number of degrees of freedom of the molecule (see Sect. 5.6; in the case of the monoatomic gas, we know already $\alpha = 3/2$). We have

$$C_V = \frac{\alpha}{2}N; \quad C_P = C_V + P\frac{\partial}{\partial T}\frac{NT}{P} = C_V + N. \quad (4.7)$$

Proceeding in the same way, we can obtain the expression for the adiabats in the PV plane, of an ideal monoatomic gas, using the relation

$$0 = \dvec{Q} = dE + PdV = (\alpha/2)d(PV) + PdV,$$

that leads to the result $C_V dP/dV = -C_P P/V$, and therefore

$$PV^\gamma = \text{const.}, \quad \gamma = C_P/C_V. \quad (4.8)$$

In general, a system that is not in thermodynamic equilibrium, will not be spatially homogeneous. Therefore, it is not possible to associate, uniquely, values of the intensive parameters P and T. A spatially homogeneous system may still be not in equilibrium, e.g. due to chemical unbalances. The increment dS will contain, in this case, a contribution from heat exchange, $= \dvec{Q}/T$, and one from chemical reactions (relaxation to chemical equilibrium): $dS^{chem} > 0$. Substituting into Eq. (4.4), and sticking with the definition of E as internal energy of the gas (chemical, plus thermal, plus possibly mechanical energy from internal motions), the first law of thermodynamics will take, in this case, the inequality form:

$$dE \leq TdS - PdV. \quad (4.9)$$

If the quasistatic hypothesis is released, e.g. due to a large temperature or pressure gap between the system and the outside world, Eq. (4.4) will cease to be valid altogether, as the action of the external world, will put the internal parts of the system in motion. We can recover some form of energy balance, in terms of simple thermodynamic quantities, if, for some reason, we can control the temperature T_0 and the pressure P_0 of the "external world". In other words, if we can consider the external world in equilibrium with itself, even if the system is not (this could be achieved e.g. with a piston determining the pressure, and a heavy metal slab fixing the temperature T_0).

In this case, the energy change ΔE is just the sum of the heat gain $-T_0 \Delta S_0$ by the system, and the work $P_0 \Delta V_0 = -P_0 \Delta V$ by the external world. Imposing the condition that the total entropy must grow, $\Delta S + \Delta S_0 \geq 0$, we obtain

$$\Delta E = -T_0 \Delta S_0 + P_0 \Delta V_0 \leq T_0 \Delta S - P_0 \Delta V. \qquad (4.10)$$

Once all the relaxation processes have terminated, we will have of course $P = P_0$ e $T = T_0$. Equation (4.10) describes therefore the relaxation of a thermodynamic system, from arbitrary initial conditions, in an external environment at fixed pressure and temperature P_0, T_0. We notice that the entropy growth, $\Delta S + \Delta S_0$, receives contribution from both internal relaxation in the system, and the fact that heat transfer occurrs between bodies (the system and the external world) at different temperature. The variation of entropy in the system is in fact a sum of different contributions

$$\Delta S(t) = \Delta S^{rel}(t) - \frac{T_0}{\bar{T}(t)} \Delta S_0(t) + \Delta S^{inho}(t),$$

where ΔS^{rel} is the contribution from a possible initial non-equilibrium condition for the system, $(T_0/\bar{T})\Delta S_0$ is the entropy variation that would be produced if the heat $-T_0 \Delta S_0$ and the work $P_0 \Delta V_0$ were generated in quasistatic manner ($\bar{T}$ is the uniform temperature that the system would have in such circumstances), and ΔS^{inho} is the contribution from inhomogeneity of the system, produced during the transient. The last term, by definition, is negative. Thus, the entropy variation $\Delta S(t)$, during the transient, will in general be smaller than what would be observed if the system transition occurred in quasistatic manner.

4.2 Conditions for a Thermodynamic Description

One of the great achievements of kinetic theory, is the microscopic derivation of the thermodynamic properties of an ideal gas. This included constitutive relations, Eqs. (3.41) and (3.50), as well as a form of the first law of thermodynamics, Eq. (3.67), and, implicitly, through the H theorem, also of the second law of thermodynamics.

We wonder under what conditions, a generic macroscopic system composed by a great number of microscopic elements will admit a thermodynamic description. From analysis of the ideal gas case, two conditions seem to play an important role:

- Distinct macroscopic parts of the system must behave, in the first approximation, as independent objects, so that energy (or some other conserved quantity with analogous role) is extensive.
- The system must tend naturally to a thermal equilibrium state, in which the system properties are maximally uniform. It must be possible to define a reasonable thermodynamic entropy, that becomes maximum at thermodynamic equilibrium. Such entropy should be itself an extensive quantity, a convex increasing function of energy.

At the microscopic scale, these conditions translate into the requirements that

- The microscopic dynamics be itself conservative, and correlations decay sufficiently fast with distance.
- Some mechanism of memory loss at microscopic scales must be present.

We have seen an example of the connection bewteen memory loss at microscopic
scales and thermodynamic behavior at macroscopic scales, in the case of the gas of
random walkers in a box (Sect. 2.7.1). In that case, thermal equilibrium of the system
(uniform density) corresponded trivially to statistical equilibrium for the walker
distribution. Unfortunately, this correspondence between relaxation to equilibrium
at microscopic and macroscopic scales, is difficult to prove for generic systems. We
shall return to these issues in the chapter on statistical mechanics.

4.2.1 Thermodynamic Inequalities

We have seen in Sect. 3.7 that the fact that entropy S of an isolated system is maximum
at thermal equilibrium, allows to define temperature as the inverse of the derivative
of the entropy with respect to the internal energy E: $T = (\partial S/\partial E)^{-1}$. This imposes
important constraints on the form of the function $S(E)$. Basically, as discussed in
Sect. 3.7, entropy must be a convex increasing function of energy. The fact that
$S(E)$ is increasing, implies obviously positive temperature: $T = (\partial S/\partial E)^{-1} > 0$.
Convexity, in turn, requires positivity of the heat capacity at constant volume. From
Eq. (4.6), we have in fact:

$$C_V = \left(\frac{\partial}{\partial E}\left(\frac{\partial S}{\partial E}\right)^{-1}\right)^{-1} = -\frac{1}{T^2}\frac{\partial^2 S}{\partial E^2} > 0 \qquad (4.11)$$

(we recall that we are considering S as a function of E and V). Let us look more
closely at the conditions $T, C_V > 0$. We can touch by hand the kind of problems that
would arise if a system had a negative temperature. Suppose we have two systems
A and B, with $T_A < 0$ and $T_B > 0$. Suppose that the condition of maximum entropy
at equilibrium continues to hold, and that thermodynamic equilibrium is still the
condition towards which A and B must tend, if put in contact. The exchanged heat
will be $\Delta Q_A = -\Delta Q_B$, with $\Delta S_A \sim \Delta Q_A/T_A$ and $\Delta S_B \sim \Delta Q_B/T_B$ (assume
the systems thermally isolated from the external world). Imposing global entropy
growth during approach to equilibrium:

$$\Delta S = \Delta S_A + \Delta S_B \sim \left(\frac{1}{T_A} - \frac{1}{T_B}\right)\Delta Q_A > 0,$$

we would find however $\Delta Q_A < 0$: heat flows from the body at negative temperature,
to the one at positive temperature, whatever the temperature of the second. A body
at negative temperature would behave as if it were hotter than any body at $T > 0$.
Negative temperatures are "hotter" than infinity.

Fig. 4.1 Entropy profile for two identical systems with $C_V < 0$, as function of their internal energy. The less energetic system, A, continues to lose energy to the more energetic one, B, until the energy content of the first is zero

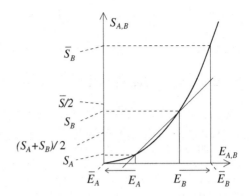

Not less paradoxical would be the behavior of a system characterized by a negative specific heat. Such a system would react to external heating with a decrease in its temperature. It is interesting to see what would happen if we put two such systems in contact. Suppose the systems identical. The situation is illustrated in Fig. 4.1: the curve $S_{A,B}$ has the wrong convexity: its points lie below the chord connecting the initial conditions for the two systems A and B. Making the system spatially uniform would lead to a decrease of the global entropy $S = S_A + S_B$. Entropy becomes maximum for $\bar{E}_A = 0$ and $\bar{E}_B = E_A + E_B$.

Similar reasoning can be carried out with pressure, that plays with respect to mechanical equilibrium, the same role played by temperature with respect to thermal equilibrium. Like temperature, pressure must be positive: as a body with negative temperature would continue to loose heat to the environment, never reaching (as long as $T < 0$)) thermal equilibrium, a body with negative pressure would simply collapse. Again, energy would be released in the environment and mechanical equilibrium could not be reached (also in this case, as long as $P < 0$). In order for mechanical equilibrium to be possible, it is necessary that pressure be positive, and, in order for equilibrium to be stable, pressure must increase in response to compression. In other words, the compressibility of the system must be positive:

$$-\frac{1}{V}\left(\frac{\partial V}{\partial P}\right)_T > 0. \tag{4.12}$$

If this condition were not satisfied, a body in mechanical equilibrium with the environment would react to an increase in the ambient pressure with a decrease of internal pressure. The resulting unbalance would produce even stronger compression of the body, and further decrease of internal pressure, in a runaway process leading again to body collapse.

The two relations (4.11) and (4.12) are called **thermodynamic inequalities**. Notice that in the two cases we have worked at constant V and T, respectively. This choice is justified, a posteriori, in the case of an ideal case, where we can verify that $C_P > C_V$ and $-(\partial P/\partial V)_S > -(\partial P/\partial V)_T$. In reality, it is possible to show that such ordering of $C_{V,P}$ and $-(\partial P/\partial V)_{S,T}$, is satisfied in general. Thus,

Eqs. (4.11) and (4.12) are sufficient conditions for stability of the system. A more formal derivation allows to verify this point.

4.2.2 Formal Derivation

Imagine that the system is part of an environment at fixed pressure and temperature. The composite system formed by the system and the environment is supposed isolated and in thermodynamic, as well as in mechanical equilibrium. Suppose now that a fluctuation occurs in the system (such fluctuations, due to the microscopic nature of the system will always be present, although asymptotically small in the thermodynamic limit). Stability would require that any such fluctuation should die away.

As a result of the fluctuation, some energy will be exchanged, in the form of heat and work, between the system and the environment. Indicating with subscript zero the environment, the total change of internal energy in the system and the environment will be, to linear order in δV_0 and δS_0:

$$\delta E^{tot} = \delta E + \delta E_0 = \delta E + T_0 \delta S_0 - P_0 \delta V_0.$$

At equilibrium, $T_0 = T$, $P_0 = P$, and the transformation occurs reversibly: $\delta S + \delta S_0 = 0$. Hence, exploiting also the fact that the total volume $V + V_0$ remains constant:

$$\delta E^{tot} = \delta E - T \delta S + P \delta V, \tag{4.13}$$

that tells us, from the first law of thermodynamics Eq. (4.4), that δE^{tot} is quadratic in δS and δV. As the environment is supposed much bigger than the system, the quadratic terms come from the system (such terms account for the change of pressure and temperature that occur in system):

$$\delta E^{tot} = \frac{\partial^2 E}{\partial V^2} \delta V^2 + 2 \frac{\partial^2 E}{\partial V \partial S} \delta V \delta S + \frac{\partial^2 E}{\partial S^2} \delta S^2.$$

The quantity $-\delta E^{tot}$ is the energy available to generate internal motions, that could be used to push the composite system farther from equilibrium. For stability, we must have $\delta E^{tot} > 0$. This requires that the following two conditions be satisfied:

$$\frac{\partial^2 E}{\partial S^2} > 0; \qquad \frac{\partial^2 E}{\partial S^2} \frac{\partial^2 E}{\partial V^2} - \left(\frac{\partial^2 E}{\partial V \partial S} \right)^2 > 0. \tag{4.14}$$

We recognize in the first condition, the first thermodynamic inequality Eq. (4.11). To obtain the second thermodynamic inequality, it is convenient to express the second of Eq. (4.14) in terms of Jacobian determinants. We can write

$$\frac{\partial^2 E}{\partial S^2} \frac{\partial^2 E}{\partial V^2} - \left(\frac{\partial^2 E}{\partial V \partial S} \right)^2 \equiv \frac{\partial \left(\frac{\partial E}{\partial S}, \frac{\partial E}{\partial V} \right)}{\partial (V, S)} = -\frac{\partial (T, P)}{\partial (S, V)} > 0.$$

Changing variables, $S, V \rightarrow T, V$, and using the properties of the Jacobian, we obtain:

$$\frac{\partial(T, P)}{\partial(S, V)} = \frac{\frac{\partial(T, P)}{\partial(T, V)}}{\frac{\partial(S, V)}{\partial(T, V)}} = \frac{\left(\frac{\partial P}{\partial V}\right)_T}{\left(\frac{\partial S}{\partial T}\right)_V} = \frac{T}{C_V}\left(\frac{\partial P}{\partial V}\right)_T < 0, \qquad (4.15)$$

that is Eq. (4.12).

4.2.3 Le Chatelier Principle

The thermodynamic inequalities, Eqs. (4.11) and (4.12), are a manifestation of the stability of thermodynamic equilibrium. This stability condition can be seen as a tendency of the thermodynamic system to reorganize in response to external perturbations, in such a way that the modification of its macroscopic parameters are minimal. This property becomes more striking in the presence of internal degrees of freedom in the system, such as chemical composition, possibility of different phases, magnetization, etc.

An example is provided by a mixture of chemicals, that can transform from one to another through reactions that will be exothermic in one sense, endothermic in the other. The system will react to external heating by rising its temperature. In response to the temperature change, the chemical equilibrium between substances will change. It is easy to be convinced that the system equilibrium will be stable, only if the reactions leading to the new chemical equilibrium, are endothermic, so that the final temperature will be lower than in the absence of reactions. (If this were not the case, the system temperature would rise even more, leading to further rise in the reaction rate in a runaway process). We see that the chemical reactions act to decrease the perturbation produced in the system.

This is the content of the **Le Chatelier principle**:

- The internal processes generated in a thermodynamic system by an external perturbation, will tend to minimize the response of the system to the external perturbation.

This tendency towards "homeostasis" can be expressed in formal way. Identify in a system an internal parameter y, and an external one x that describes the interaction of the system with the environment. The system could be e.g. a mixture of gases, with x the volume, and y the relative concentration of a certain chemical. Indicate with S the total entropy of the system and the environment, considered as an isolated composite system. The two conditions of equilibrium of the system with the environment, and internal equilibrium, are encoded in the relations

$$X = -\frac{\partial S}{\partial x} = 0, \quad \text{and} \quad Y = -\frac{\partial S}{\partial y} = 0. \qquad (4.16)$$

Stability of equilibrium will require, in analogy with Eq. (4.14):

$$\left(\frac{\partial X}{\partial x}\right)_y > 0, \quad \left(\frac{\partial Y}{\partial y}\right)_x > 0 \tag{4.17}$$

and

$$\left(\frac{\partial X}{\partial x}\right)_y \left(\frac{\partial Y}{\partial y}\right)_x - \left(\frac{\partial X}{\partial y}\right)_x^2 > 0. \tag{4.18}$$

Suppose now that we exert an external action on the system by modifying the parameter x, pushing in this way the system artificially out of equilibrium with respect to the environment. (In our example, we compress the gas; the perturbation is therefore the pressure in addition to that of the environment). In consequence of this, X will not be zero anymore, rather

$$\Delta X_y \simeq \left(\frac{\partial X}{\partial x}\right)_y \Delta x.$$

At the same time, the system will find itself out of equilibrium also internally. Thus, initially, also $Y \neq 0$. If we maintain the perturbation long enough, however, the system will reach a new internal equilibrium, characterized by a new value of y such that $Y = 0$ again. In response to this, the equilibrium variable X will take itself a new value

$$(\Delta X)_{Y=0} \simeq \left(\frac{\partial X}{\partial x}\right)_{Y=0} \Delta x.$$

We can express $(\partial X/\partial x)_Y$ in terms of Jacobians. Working as in the case of Eq. (4.15):

$$\left(\frac{\partial X}{\partial x}\right)_Y = \frac{\partial(X,Y)}{\partial(x,Y)} = \frac{\dfrac{\partial(X,Y)}{\partial(x,y)}}{\dfrac{\partial(x,Y)}{\partial(x,y)}} = \left(\frac{\partial X}{\partial x}\right)_y - \frac{\left(\dfrac{\partial X}{\partial y}\right)_x^2}{\left(\dfrac{\partial Y}{\partial y}\right)_x}.$$

The denominator to RHS of this equation, from the first of Eq. (4.17) is positive. Hence:

$$\left(\frac{\partial X}{\partial x}\right)_y > \left(\frac{\partial X}{\partial x}\right)_{Y=0} > 0,$$

where the second inequality descends from Eq. (4.18). We have finally

$$|(\Delta X)_y| > |(\Delta X)_{Y=0}|.$$

The internal modifications, produced in the system, lead to a weaker response to the external perturbation.

4.3 Thermodynamics: Macroscopic Point of View

The nice thing about thermodynamics, is that it can be derived without any information on the microscopic structure of the system. This derivation is carried out in a precise sequence of steps:

- A thermodynamic system will be described by a certain number of variables with a direct mechanical interpretation, such as volume and pressure.
- Energy conservation is enforced through the first law of thermodynamics,

$$dE = đQ - đW = đQ - PdV,$$

  that is based on the identification of an internal energy E, that can be exchanged either in the form of work W or heat Q. The internal energy of a system is a **function of state**, i.e. a function of the thermodynamic variables of the system.
- A properly defined concept of thermodynamic equilibrium must exist, which requires that the so called **zero principle of thermodynamics** must be satisfied: if two bodies A and B are in equilibrium with a third one, then A and B are in equilibrium among themselves. This allows to identify temperature as a parameter characterizing systems mutually in equilibrium.
- Thermodynamic systems have a tendency towards thermal equilibrium through heat exchange. This allows to define temperature in such a way that heat flows always from systems at higher temperatures to systems at lower temperatures. This is basically the **second law of thermodynamics** in its **Clausius** form: *physical processes, whose only result is the heat transfer from a cold body to a hot body, do not exist in nature.*
- Once a temperature scale is established, the temperature of a system can be expressed as a function of the thermodynamic variables of that system, by means of a law of state. Notice that, in this framework, the only point in which the structure of the system plays a role, is the expression of the internal energy, and the law of state.
- A thermodynamic entropy S can finally be introduced, playing with respect to temperature and heat exchange, a role similar to that of volume with respect to pressure and work:

$$dS = \frac{đQ}{T}. \tag{4.19}$$

This definition satisfies the condition of entropy growth in the approach to equilibium, and leads to the symmetric form of the first law of thermodynamics, given in Eq. (4.4).

This is the classical procedure for the derivation of thermodynamics. Its main difficulty is that nothing guarantees that the entropy, defined in Eq. (4.19), actually exists. An alternative axiomatic approach, that bypasses this difficulty, is to assume from the start the existence of an extensive entropy, and to express the law of state, and the internal energy, in terms of the entropy, rather than of the temperature of the system. This is basically the situation in which one ends up when deriving the thermodynamics of a system from its microscopic properties.

If we follow the standard procedure, however, we had better to verify that Eq. (4.19) makes sense.

Let us start by verifying this condition in the case of an ideal gas. The condition that S is a function of state, is equivalent to stating that the dS in Eq. (4.19) is exact. In other words, we must verify that there exist a function $S(V, T)$ such that

$$dS = \frac{\partial S}{\partial T}dT + \frac{\partial S}{\partial V}dV.$$

On the other hand, from the first law of thermodynamics, Eq. (4.4), we have, writing all terms as functions of V and T:

$$dS = \frac{1}{T}(dE + PdV) = \frac{1}{T}\left[\frac{\partial E}{\partial T}dT + \left(\frac{\partial E}{\partial V} + P\right)dV\right]. \qquad (4.20)$$

Using the two relations $P(V, T) = NT/V$ e $E(V, T) = (3/2)NT$, we obtain:

$$dS = \frac{3N}{2T}dT + \frac{N}{V}dV.$$

The condition that dS be exact imposes that the cross derivatives $\partial_V \partial_T S$ and $\partial_T \partial_V S$ be identical. We find in fact:

$$\frac{\partial^2 S}{\partial V \partial T} = \frac{\partial}{\partial V}\frac{3N}{T} = 0, \qquad \frac{\partial^2 S}{\partial T \partial V} = \frac{\partial}{\partial T}\frac{N}{V} = 0,$$

and recover the result that the entropy of an ideal gas exists, as expected.

In the general case, the existence of an entropy function is not guaranteed at all. Existence of entropy as a state function, imposes precise conditions on the possible form of the law of state and the internal energy of the system. We see in fact that the condition that dS be exact, imposed on the RHS of Eq. (4.20):

$$\frac{\partial}{\partial V}\frac{1}{T}\frac{\partial E}{\partial T} = \frac{\partial}{\partial T}\left(\frac{\partial E}{\partial V} + P\right),$$

leads to the constraint:

$$\frac{\partial E}{\partial V} = T\frac{\partial P}{\partial T} - P.$$

Not all expressions $U = U(V, T)$ and $P = P(V, T)$ guarantee existence of a entropy as a function of state: $S = S(V, T)$.

We conclude by calculating explicitly the thermodynamic entropy of an ideal gas. The integral $\int_{T_0, V_0}^{T, V} dS$ does not depend on the path in the plane T, V along which it is calculated, and can be decomposed therefore in the two pieces $(T_0, V_0) \to (T, V_0)$ and $(T, V_0) \to (T, V)$:

$$S(T, V) = S(T_0, V_0) + \int_{T_0}^{T} dS(T', V_0) + \int_{V_0}^{V} dS(T, V').$$

Writing $T dS = dE + P dV = (3/2) N dT + (NT/V) dV$:

$$S(T, V) = S(T_0, V_0) + N \left[\frac{3}{2} \int_{T_0}^{T} \frac{dT'}{T'} + \int_{V_0}^{V} \frac{dV'}{V'} \right]$$

$$= S(T_0, V_0) + N \left(\frac{3}{2} \ln T/T_0 + \ln V/V_0 \right),$$

that implies

$$S(T, V, N) = N \left(\frac{3}{2} \ln T + \ln V \right) + C(N).$$

The form of $C(N)$ is determined imposing the extensive property for S, i.e., that given two systems A and B: $S_{A \cup B} = S_A + S_B$. This gives us back the kinetic theory result in Eq. (3.33), obtained from the Maxwell-Boltzmann distribution, Eq. (3.32):

$$S(T, V, N) = N \left(\frac{3}{2} \ln T + \ln(V/N) \right) + N\zeta.$$

Note The terms in the above expression are not well defined dimensionally. We can extract from the constant ζ, the terms required to make the argument of the logarithms dimensionless. This is possible in quantum mechanics, that sets the scale for the phase space volume $\Delta x \Delta v = \hbar/m$:

$$S(T, V, N) = N \ln \left(\frac{(mT)^{3/2} V}{N \hbar^3} \right) + N\zeta.$$

At this point ζ takes the dimension of an energy per degree of temperature, and will describe the internal entropy content of an individual molecule.■

4.4 Thermal Engines

The first principle of thermodynamics tells us that the internal energy of a system can be converted into work, that could be then utilized to set an external motor in motion. In principle, all of the internal energy could be converted into work. We could e.g. let a gas expand adiabatically in a cylinder, and collect its internal energy by means of a piston. More strategies for energy extraction could be envisioned.

Most thermal engine do not utilize such a linear design, rather, they exploit some **thermodynamic cycle**. What they have in common is the fact that energy extraction is accompanied by the relaxation to equilibrium of a thermodynamic system. The simplest configuration is that of an isolated system composed of three parts: a cold thermostat A; a hot thermostat B; a body C, that can exchange heat and work with both thermostats, and can exert work on an external, thermally isolated motor. The body undergoes a sequence of transformations that brings it back to its initial configuration. At the end of the cycle, the internal energy of the body E is equal to its value at the beginning of the cycle, so that, from the first principle of thermodynamics:

$$0 = \Delta E = \Delta Q - \Delta W_{thermostats} - \Delta W_{motor}. \tag{4.21}$$

On the contrary, the heat and the work exchanged, ΔQ and $\Delta W = \Delta W_{thermostats} - \Delta W_{motor}$, will depend on the cycle. Of course, the best configuration, from the point of view of a thermal engine, would be one in which the body exerts all of its work directly on the motor: $\Delta W = \Delta W_{motor}$. In this case, Eq. (4.21) would become

$$\Delta W = \Delta Q = \Delta Q_A - \Delta Q_B, \tag{4.22}$$

that is basically the energy that the motor is able to tap from the heat that the two thermostats exchange through the body C. We may ask what is the maximum amount of work that could be extracted in the process. We notice that the work ΔW is just the energy lost by the isolated system (thermostats A and B, and the body):

$$\Delta E_{tot} = -\Delta W. \tag{4.23}$$

At the same time, the entropy S of the isolated system will rise as the consequence of the heat transfer. We know that the energy of an isolated system is an increasing function of its entropy. Thus, the work $\Delta W = E_{tot}^i - E_{tot}^f$, with $E_{tot}^{i,f}$ the initial and final energy of the isolated system, will be a decreasing function of the final entropy S_{tot}^f. In other words, the work will be maximum for a reversible process $\Delta S_{tot} = 0$. This means that the heat between thermostats and the body must be exchanged isothermally.

This is precisely the **Carnot cycle** (see Fig. 4.2): the body receives heat isothermally from the hot thermostat B (and in the process expands, exerting work on the motor; $\alpha\beta$); then it exerts work adiabatically on the motor (adiabatic expansion) decreasing its temperature to the temperature of the cold reservoir ($\beta\gamma$); then the body transfers heat to the cold reservoir (and in the process contracts, as the motor exerts work on it; $\gamma\delta$); finally, the motor compresses adiabatically the body till it reaches again the temperature of the hot reservoir ($\delta\alpha$).

We can define the efficiency of the conversion process as the ratio of the work extracted and the heat loss by the hot reservoir:

$$\eta = \frac{\Delta W}{-\Delta Q_B}. \tag{4.24}$$

Fig. 4.2 Sketch of a Carnot cycle $\alpha\beta\gamma\delta$. Continuous lines indicate isotherms; *dashed lines* adiabats. Notice that adiabats never cross the $T = 0$ isotherm

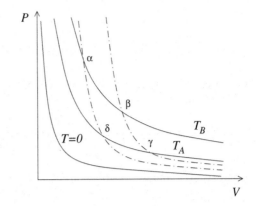

As it has already been said, the maximum is achieved in reversible conditions

$$\Delta Q_{A,B} = T_{A,B}\Delta S_{A,B}, \qquad \Delta S_A + \Delta S_B = 0. \tag{4.25}$$

(since it is a function of state, the system entropy must have identical values at the beginning and the end of the cycle). Substituting into Eq. (4.22), we find the maximum work that can be extracted from the system

$$\Delta W^{MAX} = -(T_B - T_A)\Delta S_B. \tag{4.26}$$

Dividing by the heat lost by B, we find the maximum efficiency

$$\eta^{MAX} = \frac{\Delta W^{MAX}}{-\Delta Q_B} = 1 - \frac{T_A}{T_B}. \tag{4.27}$$

We see that the efficiency of a cyclic engine goes to zero when the temperature gap between thermostats is zero. This is consistent with the **second law of thermodynamics** in its form due to **lord Kelvin**: *transformations whose only result is a conversion of heat into work, are impossible.* Notice that this formulation assumes implicitly that we are dealing with a cyclic engine (in the expansion process, considered at the beginning of this section, the final and initial states of the system were different, and one could not speak of heat to work conversion, as the only result of the process).

The cycle could be reversed, and we would obtain a refrigerator, in which heat is transferred from the cold to the hot reservoir, at the expenses of an external work source. In this case, the expression in Eq. (4.26) becomes the minimum work that must be exerted to transfer heat between the two thermostats. The fact that this work must be positive, brings us back to the second law of thermodynamics in its Clausius form.

Note It may be interesting to see in detail what happens in irreversible conditions. This corresponds to the temperature of the body, not being equal to that of the thermostat with which it gets in contact:

$$T_A < T_A^{body} < T_B^{body} < T_B. \tag{4.28}$$

In consequence of this, total entropy will not be conserved in the heat transfer:

$$\Delta Q_A = T_A \Delta S_A = -T_A^{body} \Delta S_A^{body},$$
$$\Delta Q_B = T_B \Delta S_B = -T_B^{body} \Delta S_B^{body}, \tag{4.29}$$

where $\Delta S_{A,B}^{body}$ are the entropy variations in the body, during heat transfer with A and B, respectively. Now, as the body undergoes a cycle, we must have $\Delta S_A^{body} + \Delta S_B^{body} = 0$. Thus, substituting Eq. (4.29) into Eq. (4.24), and using Eq. (4.28), we find

$$\eta = 1 - \frac{T_A^{body}}{T_B^{body}} < 1 - \frac{T_A}{T_B}.$$

A similar efficiency reduction would be obtained if the heat between body and reservoirs were transferred reversibly, but losses were present, in the form of a direct heat transfer between A and B.■

4.5 Nernst Theorem: Third Law of Thermodynamics

We have seen that reversing the thermodynamic cycle in Fig. 4.2, turns a thermal engine into a refrigerator. While in the case of the thermal engine, a very large ratio T_B/T_A guaranteed that most of the heat lost by the hot thermostat B was converted into work, in the case of the refrigerator, a large T_B/T_A would result in most of the work being spent to heat B, and comparatively little heat being transferred from A to B. This tells us that cooling an object, that is already at a very low temperature, risks to be very energy consuming, unless we have at our disposal a "hot" thermostat that is already very cold.

This difficulty is reflected in the singularity of the zero temperature limit (the so called absolute zero). The singularity is apparent in the behavior of the entropy of an ideal gas at $T \rightarrow 0$, as can be checked either from Eq. (3.33), or integrating Eq. (4.19) in constant volume conditions:

$$S(T) = S(T_0) + C_V \int_{T_0}^{T} \frac{dT'}{T'} = S(T_0) + C_V \ln(T/T_0).$$

The thermodynamic entropy of the ideal gas diverges logarithmically in the limit $T \rightarrow 0$. We know however that this expression cannot be correct: somewhere in the passage from Shannon to thermodynamic entropy, the property that the minimum entropy of a system must be zero (and not $-\infty$), has been lost.

The fact is that expressions, such as Eq. (4.19), assume a continuous limit of something that is intrinsically granular at microscopic scales. This granular nature was motivated, in the first place, by the need for a well defined concept of entropy, which required a discrete partition of the microscopic phase space of the system (the Γ-space). Quantum mechanics made this requirement more substantial, introducing a fundamental volume scale in phase space, in terms of the Planck constant $\hbar$.

If we maintain our definition of temperature as a proxy for thermodynamic equilibrium, which is the content of the expression

$$T = \left(\frac{\partial E}{\partial S}\right)_V,$$
(4.30)

we expect that the minimum of the entropy, $S = 0$, should be achieved at zero temperature. This statement can in fact be made formal, and is the content of the **Nernst theorem**.

From the discussion in Sect. 3.42, we expect that quantum effects become dominant in this limit. A flavour of the Nernst theorem can nevertheless be obtained also classically (at least in the case of the ideal gas), provided we stick to the definitions of entropy in terms of a partition of Γ space, provided in Sect. 3.4.

In this picture, the velocity space of the individual molecule is discretized at a scale δv; the kinetic energy of the molecules is discretized in consequence. The zero entropy limit of the system will thus correspond to a regime, in which the probability that a molecule does not lie in the cell with the smallest possible energy, ϵ_0, is very small. To calculate the deviations from absolute zero, we could limit ourselves to consider molecules in energy levels ϵ_0 and ϵ_1, and disregard the exponentially smaller probability of finding molecules at higher energy levels.

If α is the number of cells at the next energy level ϵ_1, and p is the probability of one of the molecules to occupy one of these cells, the entropy of the system will be, from Eqs. (2.19) and (3.18):

$$S \simeq -N[(1 - \alpha p) \ln(1 - \alpha p) + \alpha p \ln p],$$
(4.31)

while the internal energy will read:

$$E \simeq N[(1 - \alpha p)\epsilon_0 + \alpha p\epsilon_1].$$
(4.32)

Substituting Eqs. (4.31) and (4.32) into Eq. (4.30), we find

$$T = \frac{\partial E}{\partial p}\left(\frac{\partial S}{\partial p}\right)^{-1} \simeq -\frac{\epsilon_1 - \epsilon_0}{\ln \alpha p} \simeq -\frac{\epsilon_1 - \epsilon_0}{\ln(S/N)},$$
(4.33)

and we verify that $S(T = 0) = 0$. In this picture, nothing dramatic takes place in the limit $T \to 0$: simply, the energy of the molecules goes below the definition at which we are studying the microscale. In particular, the system will still have a pressure associated with the residual energy ϵ_0

$$P \sim N\epsilon_0/V,$$

that remains finite also at $T = 0$. Therefore, we can identify in the PV plane, a $T = 0$ isotherm that is at the same time an adiabat at $S = 0$, and no adiabat, crossing an isotherm at $T \neq 0$, can intersect the $T = 0$ isotherm (see Fig. 4.2). This means that no finite thermodynamic cycle can involve a thermostat at $T = 0$. This is the content of the **third law of thermodynamics**: it is not possible to make a system reach zero temperature through a finite sequence of finite (reversible) thermodynamic cycles.

4.6 Thermodynamic Potentials

The first law of thermodynamics, $dE = TdS - PdV$, allows to identify the internal energy E of a system as an attitude to execute work and exchange heat with the outside world. In particular, TdS is the heat exchanged reversibly at constant volume, and $-PdV$ is the work executed isoentropically (i.e. through an adiabat). One obvious question regards the existence of state functions describing the attitude of a system to execute work or exchange heat in general conditions. The thermodynamic potentials are introduced with this goal. In particular, that of describing isobaric heat transfer and isothermal work.

As focus on isoentropic work and constant volume heat exchange forced us to consider E as a function of S and V, focus on isothermal work or isobaric heat exchange will force us to consider a thermodynamic potential that depends on T, V and S, P, respectively. It is rather clear that a systems that executes work isothermally, will require a heat bath to maintain its temperature constants. Similarly, work will have to be exerted on the system to maintain pressure constant while it exchanges heat. Thus, it is not only the system energy that is exchanged in the process, but also that of the thermal bath, in the first case, and of some piston, in the second. In short, the system is not isolated: the attitude to execute work or exchange heat is not "contained" only in E. It will turn out that the additional contribution to energy can be obtained by means of a Legendre transformation, that is precisely the transformation producing the change of variables $S \to T$ in one case, $V \to P$ in the other.

4.6.1 Enthalpy

The first potential that we introduce, describes the attitude to exchange heat at constant pressure. It is called **enthalpy**:

$$H(S, P) = E(S, V(S, P)) + PV(S, P), \qquad dH = TdS + VdP. \qquad (4.34)$$

The second relation in Eq. (4.34), is valid only in reversible conditions, meaning that no internal relaxation processes are supposed to take place; the only transformations in the system are heat transfer and compression. The volume $V(S, P)$ is defined through the condition that $E(S, V) + PV$ be minimum with respect to V, i.e.:

$$P(S, V) = -\frac{\partial E(S, V)}{\partial V}.$$ (4.35)

Mathematically, the mechanism for the change of variables $S(S, V) \to (S, P)$, in Eqs. (4.34) and (4.35), is a Legendre transform. From the point of view of physics, we see that enthalpy is the sum of the internal energy of the system and the energy of a piston that makes the volume of the system take the value $V(S, P)$, such that the internal pressure of the system balances the pressure P of the piston. In order for pressure to remain constant, it is in fact necessary that the environment executes some word $P \Delta V$ on the system, as illustrated in Fig. 4.3.

The variation of H at constant pressure coincides with the heat exchanged, as required: $dH = dE + d(PV) = dE + PdV$. This allows us to write the heat capacity at constant pressure, in terms of the enthalpy:

$$C_P = \left(\frac{\partial H}{\partial T}\right)_P$$ (4.36)

(compare with Eq. (4.6)).

Once an explicit expression for the enthalpy $H(S, P)$ is available, we can write the volume $V(S, P)$ as function of the pressure:

$$V(S, P) = \frac{\partial H(S, P)}{\partial P}.$$ (4.37)

In similar way, the temperature can be expressed as a function of the entropy and the pressure by means of the relation

$$T(S, P) = \frac{\partial H(S, P)}{\partial S}.$$ (4.38)

Fig. 4.3 Geometric construction of the Legendre transform for the passage from energy to enthalpy

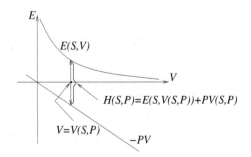

Fig. 4.4 The Joule-Thomson
apparatus

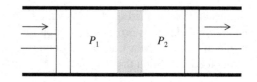

4.6.2 Joule-Thomson Process

The concept of enthalpy can be used to study the energetics of the so called Joule-Thomson process, that is the quasistatic flow of a gas (or a liquid) through a porous medium, induced by a pressure gap. The experimental setting is sketched in Fig. 4.4: the gas is contained in a cylinder divided in two sectors by a porous slab. A constant pressure gap is induced between the two sectors by a couple of pistons. Both the pistons and the cylinder walls are thermal insulators.

Suppose that, due to a pressure gap $P_1 > P_2$, the volume of gas V_1 to the left of the slab flows to the right, where, at the end of the process, it will occupy volume V_2. Since the gas does not exchange heat with the environment, the only change in its internal energy is the one produced by the work of the pistons:

$$E_2 - E_1 = P_1 V_1 - P_2 V_2.$$

In other words:

$$H_1 = E_1 + P_1 V_1 = E_2 + P_2 V_2 = H_2;$$

the enthalpy of the gas is conserved in the process.

We can use this information to calculate important parameters, e.g. the temperature change produced in the gas by the process. For a small pressure gap, $\Delta P = P_2 - P_1 << P_{1,2}$, we could approximate

$$\Delta T = \left(\frac{\partial T}{\partial P}\right)_H \Delta P,$$

where

$$\left(\frac{\partial T}{\partial P}\right)_H = \frac{\partial(T, H)}{\partial(P, H)} = \frac{\frac{\partial(T, H)}{\partial(P, T)}}{\frac{\partial(P, H)}{\partial(P, T)}} = -\frac{\left(\frac{\partial H}{\partial P}\right)_T}{\left(\frac{\partial H}{\partial T}\right)_P}. \tag{4.39}$$

Exploiting Eqs. (4.36) and (4.37), Eq. (4.39) can be rewritten

$$\left(\frac{\partial T}{\partial P}\right)_H = \frac{1}{C_P}\left[T\left(\frac{\partial V}{\partial T}\right)_P - V\right]. \tag{4.40}$$

In the case of an ideal monoatomic gas, for which $V = NT/P$, we would find $(\partial V/\partial T)_P = 0$, and therefore $\Delta T = 0$. This is to be contrasted with the result in an adiabatic transformation, for which, combining Eq. (4.8) with the law of state $PV = NT$, we would have $dT = \frac{2T}{5P}dP$. Although both transformations occur without exchange of heat, moving along the adiabat of Eq. (4.8) corresponds to a reversible (isoentropic) transformation, while the Joule-Thomson process is irreversible and involves an increase of entropy in the gas.

4.6.3 Helmholtz Free Energy

The next thermodynamic potential that we are going to introduce, called the Helmholtz free energy, describes the attitude of the system to execute work isothermally:

$$F(T, V) = E(S(T, V), V) - TS(T, V), \qquad dF = -SdT - PdV. \qquad (4.41)$$

As in Eq. (4.34), the second of Eq. (4.41) is valid only in reversible conditions. We easily verify that the variation of free energy at constant temperature, $dF = dE - d(TS) = -PdV$, is just the work executed on the system, as required.

In a way analogous to enthalpy, the dependency of entropy on the variables (T, V) is determined through a Legendre transform of E with respect to S. In this case, the minimum condition is imposed on $E(S, T) - TS$. This gives us $S(T, V) = \partial E(S, V)/\partial S$, which can be inverted to produce $S = S(T, V)$. Once the explicit expression for $F(T, V)$ is known, we can utilize the second in Eq. (4.41) to determine $S(T, V)$:

$$S(T, V) = -\frac{\partial F(T, V)}{\partial T}. \qquad (4.42)$$

Again, the system executes work on some other object, while the environment acts as a thermal bath, and provides the necessary heat to maintain the system isothermal. Once the $S(T, V)$ is known, we can determine the expression of the pressure as function of the temperature and the volume of the system

$$P(T, V) = -\frac{\partial F(T, V)}{\partial V}, \qquad (4.43)$$

that provides the law of state of the system.

Irreversible processes, going on in the interior of the system, will produce variations in its free energy, not accounted for by the second of Eq. (4.41). In particular, the work at constant temperature will not equal the change of the Helmholtz free energy at constant temperature. From the inequality form of the first law of thermodynamics, Eq. (4.9), we have in fact

$$dF = d(E - TS) \leq TdS - PdV - d(TS) = -SdT - PdV. \qquad (4.44)$$

In particular, for a process taking place at constant volume and temperature, we find

$$dF \leq 0. \tag{4.45}$$

The Helmholtz free energy of a system at constant T and V becomes minimum at thermodynamic equilibrium.

4.6.4 Gibbs Free Energy

Taking the Legendre transform of the enthalpy with respect to entropy, or alternatively, of the Helmholtz free energy with respect to volume, we obtain still another thermodynamic potential, the Gibbs free energy, that will be a function of T and P:

$$\Phi(T, P) = E - TS(T, P) + PV(T, P), \quad d\Phi = -SdT + VdP. \tag{4.46}$$

Again, as in Eqs. (4.34) and (4.41), the second of Eq. (4.46) is valid only in reversible conditions. As we have said, the Gibbs free energy can be obtained either as the Legendre transform of the enthalpy, with respect to S, or of the Helmholtz free energy, with respect to V. The functions $S(T, P)$ and $V(T, P)$ are obtained from inversion of $T = \partial H(S, P)/\partial S$ and $P = -\partial F(T, V)/\partial V$, respectively. Once the explicit form of $\Phi(T, P)$ is known, the volume and the entropy of the system can be determined from

$$S(T, P) = -\frac{\partial \Phi(T, P)}{\partial T} \quad \text{and} \quad V(T, P) = \frac{\partial \Phi(T, P)}{\partial P}. \tag{4.47}$$

Contrary to the case of the enthalpy and of the Helmholtz free energy, the Gibbs free energy is not easily interpretable as an attitude to reversibly execute work or exchange heat. As it will be discussed in the next section, the importance of the Gibbs free energy lies in its ability to describe the attitude of a system, undergoing internal relaxation processes (e.g. chemical reactions), to execute work at constant pressure and temperature. Let us then examine the modifications that must be introduced in the second of Eq. (4.46) in irreversible conditions. Proceeding as in the case of the Helmholtz free energy, we find

$$d\Phi = d(E + PV - TS) \leq TdS - PdV + d(PV - TS) = -SdT + VdP, \tag{4.48}$$

that, at constant pressure and temperature, gives us

$$d\Phi \leq 0. \tag{4.49}$$

The Gibbs free energy in systems maintained at constant pressure and temperature, becomes minimum at thermal equilibrium.

4.7 Free Energy as an Internal Source of Work

We want to understand more in detail in which sense the free energy can be interpreted as the attitude to execute work of a non thermally isolated system. Let us proceed in a way similar to Sect. 4.4 and consider a composite macrosystem composed by the environment and a system that can execute work on a thermally isolated motor. We make the following assumptions:

- The total volume of the macrosystem remains constant, so that changes of volume in the system and in the environment must be equal and opposite: $\Delta V = -\Delta V_0$ (as in Sect. 4.1, we shall identify variables referred to the environment with subscript 0).
- The motor interacts only with the system, through execution of mechanical work.
- The system can execute work and exchange heat with the environment.
- The environment is very large, so that the transformations taking place in the system do not modify its state (the environment is a thermal and pressure bath).

Let us start by calculating the change in internal energy of the system, during a transformation involving a work $P_0 \Delta V_0$ from the environment, a heat transfer $-T_0 \Delta S_0$ again from the environment, and a work ΔW_{in} executed by the motor. We find immediately, exploiting $\Delta V_0 = -\Delta V$ and $\Delta S + \Delta S_0 \geq 0$:

$$dE = dW_{in} - T_0 dS_0 + P_0 dV_0 \leq dW_{in} + T_0 dS - P_0 dV,$$

i.e.

$$dW_{in} \geq dE - T_0 dS + P_0 dV. \tag{4.50}$$

This expression tells us that the work that the motor must execute on the system to induce the transformation $\{\Delta E, \Delta S, \Delta V\}$, will be minimum in reversible conditions: $\Delta S + \Delta S_0 = 0$. On the contrary, the work that the system can execute on the motor during a transformation $\{\Delta E, \Delta S, \Delta V\}$, will always be smaller than $-(\Delta E - T_0 \Delta S + P_0 \Delta V)$. We find therefore that the maximum amount of work that the system can execute during the transformation, is the reversible work

$$dW_{out}^{MAX} = -(dE - T_0 dS + P_0 dV). \tag{4.51}$$

Within a change of sign, we recognize in the RHS of the above formula, the differential of the "precursor" Gibbs free energy $\hat{\Phi}_{T_0 P_0}(S, V)$, whose minimum with respect to S and V is precisely $\Phi(T_0, P_0)$. Likewise, if we considered a constant volume process, $dV = 0$, we would have $dW_{out}^{MAX} = -(dE - T_0 dS) = -d\hat{F}_{T_0}(S, V)$, with $\hat{F}_{T_0}$ the precursor Helmholtz free energy, whose minimum with respect to S is precisely $F(T_0, V)$.

In the absence of internal relaxation processes, we can use the first law of thermodynamics, Eq. (4.4), to write the variation of internal energy of the system as $\Delta E = P \Delta V - T \Delta S$, and therefore, exploiting Eq. (4.51):

$$\mathrm{d}W_{out}^{MAX} = (T_0 - T)\mathrm{d}S - (P_0 - P)\mathrm{d}V.$$

In the absence of internal energy sources (chemical or others), the only mechanism of work generation, is unbalance between the system and the environment.

More interesting the situation in which there is an internal energy source, in the form of an internal non-equilibrium condition of the system. This is the situation in which the free energies, introduced in the last section, come to help. We consider separately the two situations in which the system undergoes a cyclic isothermal transformation (so that $\Delta V = 0$ at the end of the cycle), and of a system executing work at fixed pressure and temperature.

In the first case, if $T = T_0$, we see immediately that ΔW_{out}^{MAX} equals the variation of Helmholtz free energy of the system. Indeed, combining Eq. (4.41) with the two conditions $\Delta V = 0$ and $T = T_0 = $ const., and substituting into Eq. (4.51), we find

$$\Delta W_{out} \le -(\Delta E - T_0 \Delta S + P_0 \Delta V) = -\Delta(E - TS) = -\Delta F. \qquad (4.52)$$

Proceeding in similar way, if the system executes work at ambient pressure and temperature ($P = P_0$ and $T = T_0$), we obtain from Eq. (4.46):

$$\Delta W_{out} \le -(\Delta E - T_0 \Delta S + P_0 \Delta V) = -\Delta(E - TS + PV) = -\Delta \Phi. \qquad (4.53)$$

The maximum work is given, in this case, by the change in the Gibbs free energy. In the two cases, the maximum work would be generated if the entropy produced in the internal process, ΔS^{int}, were fully balanced by a cooling off of the environment, with a decrease of entropy $\Delta S_0 = -\Delta S = -\Delta S^{int}$.

4.8 Systems with Varying Number of Particles

Until now, we have not considered explicitly processes involving changes in the number of molecules. There are important situations, however, in which the number of molecules does change. We shall consider in detail two examples:

- Chemical reactions that cause the transmutation of molecules from one species to another.
- Phase transition that cause the passage of molecules of the same substance from one phase to another.

During such transmutations, the internal energy of the molecules becomes available for conversion into heat (thermal motion) or mechanical work. It is then necessary that we start to take into account this energy component explicitly. In the case of a mixture of ideal gases at fixed temperature, we would have, e.g.

$$E(T, V, \{N_i\}) = \sum_i N_i \left(\frac{k_i}{2} + \epsilon_i(T) \right), \qquad (4.54)$$

where N_i, ϵ_i, and k_i are the number of molecules of the different species, their internal energies and number of degrees of freedom ($k = 3$ in the case of a monoatomic component). In the following, we shall assume that the internal energy of the molecules is independent of temperature, but it is obvious that this is only an approximation (molecular collisions may well activate internal degrees of freedom of the molecules, such as those associated with vibrations).

We thus see that a transformation that leads to an increase of the number of molecules with lower ϵ, at the expenses of the number of those with higher ϵ, will lead to a decrease of the component $\sum_i N_i \epsilon_i(T)$ in E. If the system is isolated, it is necessary that the thermal energy component increases. Thus, temperature must increase, and we have an example of an **exothermic** reaction. On the other hand, a reaction in which there is an increase of molecules with high ϵ, will require a decrease of the energetic component due to thermal motion. We speak in this case of an **endothermic** reaction.

Through the increase of thermal motion, an exothermic reaction will lead to a positive contribution to the system entropy. The growth of entropy in an exothermic reaction would suggest that the reaction should proceed to exhaustion of the reactants in the mixture. Things however do not work this way neither in an isolated system, nor in a system in contact with a thermal bath. In fact, a contribution to the system entropy is produced also by the concentration of the different species (the term $\sum_i N_i \ln V/N_i$ in the expression of the entropy of a mixture; see next section), and by the internal entropy of the molecules.

A similar phenomenon occurs in the case of a liquid-gas phase transition: the energy of a molecule in the liquid phase is less than in the gaseous phase. Thus, the liquid-gas transition releases a certain amount of heat (the so-called **latent heat**). In the isolated system, the heating would lead to a positive contribution to the entropy; nevertheless, the transition will take place only if it is allowed by the global entropy balance, that takes into account the contribution from the volume densities of the different phases. Thus, it is necessary to calculate the maximum of entropy $S(E, V, \{N_i\})$, for constant E and V. Similarly in the case in which the system is not isolated, and we proceed at constant T, V, or constant T, P, in which case, we must consider the minimum with respect to $\{N_i\}$ of the two free energies $F(T, V, \{N_i\})$ and $\Phi(T, P, \{N_i\})$.

4.8.1 Chemical Potential

In order to proceed, we must convert an equation, such as (4.54), into a form in which the internal energy depends on entropy, such as Eq. (4.4).

$$E = E(S, V, \{N_i\}); \quad dE = TdS - PdV + \sum_i \mu_i dN_i. \quad (4.55)$$

We see that a new intensive parameter, coupled with the extensive parameters $\{N_i\}$, enters the game; the **chemical potential**:

$$\mu_i = \mu_i(s, v, \{c_j\}) = \frac{\partial E(S, V, \{N_j\})}{\partial N_i} \qquad s = S/N, \quad v = V/N, \quad c_j = N_j/N,$$

$$(4.56)$$

where we have been keen to express the intensive quantity μ_i as function of other intensive parameters s, v and c_i.

Equation (4.55) is in the form of a first law of thermodynamics, in which the contribution from the internal degrees of freedom of the molecules is taken into consideration explicitly. This has the consequence that an equality form of this law is recovered, from the inequality form of Eq. (4.9).[1] In similar way, we can recover differential definitions for the free energies in equality form, from the inequalities in Eqs. (4.45) and (4.49).

The dependence of the chemical potential μ_i on the concentrations $\{c_j\}$, is important in the case of a gas mixture, and becomes essential in the determination of chemical equilibrium. In the case of phase transitions, this dependence is clearly absent: the concentration of molecules in the liquid phase in a region where the system is in gaseous phase, is of course zero, and vice versa in the part in liquid phase.

The chemical potential μ_i is the energy variation produced by the change dN_i in the number of molecules of species i, at fixed values of the volume and the entropy of the system. Notice that the constant entropy condition means that some heat must be provided to the system to compensate the entropy change in the reaction (for the sake of definiteness, let us stick for the moment to the case of chemical reactions). Thus, μ_i could not be interpreted merely as an energy contribution from the chemical bonds. In other words, in general $\mu_i \neq \epsilon_i$. As we shall see, μ_i receives contributions both from the thermal state (pressure and temperature) of the system, and from the chemical degrees of freedom (energy and entropy of the chemical bonds).

4.8.2 Helmholtz and Gibbs Free Energy

Legendre transforming Eq. (4.55) with respect to S, we obtain the Helmholtz free energy

$$F = E - TS = F(T, V, \{N_i\}), \qquad dF = -SdT - PdV + \sum_i \mu_i dN_i, \quad (4.57)$$

where now

$$\mu_i = \mu_i(T, v, \{c_i\}) = \frac{\partial F(V, P, \{N_i\})}{\partial N_i}, \qquad (4.58)$$

---

[1] Of course, in order for this to be true, it is necessary that no other irreversible processes (e.g. viscous damping of internal motions), be present in the system.

while entropy and pressure continue to obey Eqs. (4.42) and (4.43):

$$S = -\frac{\partial F}{\partial T}; \quad P = -\frac{\partial F}{\partial V}.$$

Legendre transforming Eq. (4.57) with respect to V, we obtain for the Gibbs free energy

$$\Phi = E - TS + PV = \Phi(T, P, \{N_i\}), \quad d\Phi = -SdT + VdP + \sum_i \mu_i dN_i, \quad (4.59)$$

where now

$$\mu_i = \mu_i(T, P, \{c_i\}) = \frac{\partial \Phi(T, P, \{N_i\})}{\partial N_i}, \quad (4.60)$$

while entropy and volume continue to obey Eq. (4.47):

$$S = -\frac{\partial \Phi}{\partial T}; \quad V = \frac{\partial \Phi}{\partial P}.$$

The Gibbs free energy is by construction an extensive quantity. It can therefore be written in the form

$$\Phi(T, P, \{N_i\}) = Nf(T, P, \{c_i\}),$$

i.e., Φ is homogeneous of order one with respect to the $N_i's$. Then, Eulers theorem holds:

$$\Phi(T, P, \{N_j\}) = \frac{\partial \Phi(T, P, \{\alpha N_j\})}{\partial \alpha}\bigg|_{\alpha=1} = \sum_i \frac{\partial(\alpha N_i)}{\partial \alpha} \frac{\partial \Phi(T, P, \{\alpha N_j\})}{\partial(\alpha N_i)}\bigg|_{\alpha=1}$$

$$= \sum_i N_i \frac{\partial \Phi(T, P, \{\alpha N_j\})}{\partial N_i}.$$

In other words:

$$\Phi(T, P, \{N_j\}) = \sum_i N_i \mu_i(T, P, \{c_j\}). \quad (4.61)$$

We are thus able to identify the chemical potential μ_i, as the Gibbs free energy per molecule of species i.

The minimum property at equilibrium, of the Gibbs free energy, takes now a deep physical meaning, considering that the quantity $E + PV - TS$ in the definition of Φ, is nothing but the total energy of the system and the environment, considered together as an isolated macrosystem. We could in fact consider the quantity $E - TS + PV$ as a function of the extensive parameters S, V and $\{N_i\}$:

$$\hat{\Phi}_{T,P}(S, V, \{N_i\}) = E(S, V, \{N_i\}) - TS + PV, \quad (4.62)$$

with T and P considered as external fixed parameters, and verify that thermodynamic equilibrium occurs at the minimum of the function with respect to S, V and $\{N_i\}$. If we keep the number of molecules fixed, we will recover the condition that the differential $dE - TdS + PdV$, appearing in the equilibrium condition, Eq. (4.13), is zero. The additional condition of chemical equilibrium or equilibrium between phases, is then fixed by the relation

$$\sum_i \mu_i(T, P; \{c_i\})dN_i = 0.$$

(Notice that the differentials dN_i, as it shall discussed in Sects. 4.10 and 4.11, cannot be considered independent, in one case, due to the constraints imposed by the stoichiometric ratios, in the other, simply, by conservation of the total number of molecules in the different phases).

4.8.3 Thermodynamic Potential, Function of Chemical Potential

Presence of the additional extensive parameters, $\{N_i\}$, allows to define new thermodynamic potentials, functions of the chemical potentials $\{\mu_i\}$. It is of particular relevance the thermodynamic potential obtained by Legendre transform with respect to $\{N_i\}$ of the Helmholtz free energy F:

$$\Omega = F - \sum_i \mu_i N_i, \tag{4.63}$$

where now the number of molecules of the different species are the equilibrium values, fixed by the chemical potentials, through the relations

$$\mu_i = \frac{\partial F(T, V, \{N_j\}}{\partial N_i}.$$

Now, from Eq. (4.61), we can write in Eq. (4.63):

$$\Omega = F - \Phi = -P(T, V, \{\mu_i\})V; \qquad d\Omega = -SdT - \sum_i N_i d\mu_i, \tag{4.64}$$

so that, knowledge of the functional form of the pressure $P = (T, V, \{\mu_i\})$, and the expressions for the partial derivatives

$$S = -\frac{\partial \Omega}{\partial T} \quad \text{and} \quad N_j = -\frac{\partial \Omega}{\partial \mu_j}, \tag{4.65}$$

will provide the laws of state for a system in the presence of chemical reactions.

4.9 Entropy and Thermodynamic Potentials in a Mixture of Gases

The internal degrees of freedom of a molecule can contribute to the entropy of a system, both in the form of heat release or absorption, and due to the entropic content of the molecule internal degrees of freedom. For instance, if the molecules of a certain species could be found in three equal-energy, equiprobable states, we would have an additional entropic contribution per molecule, with respect to the case in which a single energy state is possible:

$$\Delta\zeta = \sum_{\alpha=1}^{3} P(\alpha) \ln P(\alpha) = \ln 3.$$

This has the consequence that, in a transition in which ΔN single-state molecules convert into as many three-state molecules, an increment in the system entropy

$$\Delta S = \Delta N \Delta\zeta = \Delta N \ln 3$$

would occur. As in the case of the internal energy of the molecule, also the internal entropy of the molecule will be in general a function of temperature (often, higher energies are associated with an increment in the number of states accessible to the molecule). For simplicity, however, we shall limit our analysis to the case in which the internal entropy content is independent of temperature. The quantity ζ_i is called the **chemical constant** of the ith species.

We can determine the entropy of a mixture of ideal monoatomic gases, as the sum of the entropy contributions from the individual species. From Eq. (3.33):

$$S(T, V, \{N_i\}) = \frac{3}{2}NT + \sum_i N_i(\ln V/N_i + \zeta_i)$$

$$= \frac{3}{2}NT + N\ln(V/N) - N\sum_i c_i(\ln c_i - \zeta_i). \qquad (4.66)$$

We call attention to the mixing contribution $-N\sum_i c_i \ln c_i$, that is positive, and would be absent in the presence of a single substance ($c_i = 1$ for just one species; $c_i = 0$ for all the others).

At this point, we can pass to the calculation of the thermodynamic potentials. From Eq. (4.41), we obtain at once the Helmholtz free energy

$$F(T, V, \{N_i\}) = \frac{3}{2}NT + N\sum_i c_i\epsilon_i - TS(T, V, \{N_i\})$$

$$= \frac{3}{2}NT(1 - \ln T) - NT\ln(V/N) + \sum_i N_i[\epsilon_i + T(\ln c_i - \zeta_i)].$$

$$(4.67)$$

From here we can derive the law of state of a gas mixture, the so called **Dalton law**:

$$P = -\frac{\partial F}{\partial V} = \sum_i n_i T \equiv \sum_i P_i, \tag{4.68}$$

where $n_i = N_i/V$ and $P_i = n_i T$ is the **partial pressure** of the ith species. Similarly, from Eq. (4.46), we obtain for the Gibbs free energy:

$$\Phi(T, P, \{N_i\}) = F(T, NT/P, \{N_i\}) + NT$$
$$= \frac{5}{2}NT(1 - \ln T) + NT \ln P + \sum_i N_i[\epsilon_i + T(\ln c_i - \zeta_i)], \tag{4.69}$$

where use has been made of the equation of state $PV = NT$. From Eqs. (4.69) and (4.61) we can then write for the chemical potential of the i-species:

$$\mu_i = \frac{\partial \Phi}{\partial N_i} = \frac{5}{2}T(1 - \ln T) + T \ln P + \epsilon_i + T(\ln c_i - \zeta_i). \tag{4.70}$$

Notice that we can rewrite Eq. (4.70) in terms of partial pressures:

$$\mu_i = \frac{\partial \Phi}{\partial N_i} = \frac{5}{2}T(1 - \ln T) + T \ln P_i + \epsilon_i - T\zeta_i. \tag{4.71}$$

The expression for the chemical potential of the ith species in a mixture of gases, and for the same species, in pure state, differ only for the substitution $P_i \rightarrow P$. Notice at last that, differentiating Eq. (4.61), we can obtain for the specific entropies and specific volumes of the different molecules:

$$s_i = -\frac{\partial \mu_i}{\partial T} = \frac{5}{2} \ln T - \ln P + \zeta_i + \ln c_i; \qquad v_i = \frac{\partial \mu_i}{\partial P} = \frac{T}{P}. \tag{4.72}$$

From here, we recover the relation $\mu_i = u_i - Ts_i + Pv_i$, with $u_i = (3/2)T + \epsilon_i$ the mean energy of a species i molecule. Equation (4.72) reproduces at molecular scales the definition of Gibbs free energy: $\Phi = E - TS + PV$.

4.10 Chemical Reactions in a Gas Mixture

Suppose that we have a mixture of gaseous chemical species A_i, which can undergo chemical reactions characterized by the stoichiometric ratios

$$\nu_{ij} = \frac{\Delta N_i}{\Delta N_j}.$$

For instance, in the case of the reaction $2H_2 + O_2 \rightarrow 2H_2O$, we would have the ratios

$$\nu_{H_2,O_2} = 2, \quad \nu_{H_2O,O_2} = -2.$$

In the presence of a single chemical reaction, it is thus sufficient to give the variation of any of the species, to determine the variation of all the others. Taking for reference species 1, and writing $\nu_i \equiv \nu_{i1}$, we would have therefore:

$$\Delta N_i = \nu_i \Delta N_1. \tag{4.73}$$

The condition of chemical equilibrium will be given therefore by the equation

$$\frac{d\Phi}{dN_1} = \sum_i \frac{dN_i}{dN_1} \frac{\partial \Phi}{\partial N_i} = \sum_i \nu_i \mu_i = 0, \tag{4.74}$$

where use has been made of Eqs. (4.73) and (4.59). This equation allows to determine the concentrations at equilibrium of the different species.

Let us carry out the calculation in the simplest possible case of a reaction $A_1 \rightarrow A_2$ (strictly speaking this could not be a chemical reaction, as it does not involve either combination or decomposition of molecules). In this case, we would have $\nu_1 = 1$, $\nu_2 = -1$, and the equilibrium condition would be simply $\mu_1 = \mu_2$. Let us consider a reaction that takes place at constant P and T. Substituting the expression for the chemical potential, derived in the previous section, Eq. (4.70), we find at once

$$\mu_1 - \mu_2 = \epsilon_1 - \epsilon_2 + T\left(\ln(c_1/c_2) - \zeta_1 + \zeta_2\right) = 0,$$

from which we obtain

$$\frac{c_1}{c_2} = \exp\left(\zeta_1 - \zeta_2 - \frac{\epsilon_1 - \epsilon_2}{T}\right). \tag{4.75}$$

Notice the analogy with the Maxwell-Boltzmann distribution: the concentrations are weighed by an exponential of the ratio of the chemical energy of the molecules, and the temperature. We find however a new entropic contribution, that comes from the chemical constants. A reaction hindered by the fact that the energy of the reactants is smaller than that of the products of reaction, may therefore still take place, if it is advantageous from the entropic point of view, i.e. if the difference $\zeta_2 - \zeta_1$ is sufficiently large.

4.10.1 Reaction Enthalpy

The heat of reaction, that, in the isobaric conditions considered, is a reaction enthalpy, can at this point be calculated, exploiting the relation

$$H(S(T,P), P) = \Phi(P,T) + TS(T,P) = \Phi(P,T) - T\frac{\partial \Phi(T,P)}{\partial T}$$

$$= -T^2 \frac{\partial}{\partial T} \frac{\Phi(P,T)}{T},$$

where use has been made of Eqs. (4.34) and (4.46), and, to keep notation light, we have not indicated the dependence on $\{N_i\}$. The enthalpy production in the transformation of $\Delta N_2 = -\Delta N_1 > 0$ molecules of species A_1 to A_2, will be, for $\Delta N_2 \ll N$:

$$\Delta H \simeq \Delta N_1 \frac{dH}{dN_1} = \Delta N_2 T^2 \frac{\partial}{\partial T}\left(\frac{1}{T}\frac{d\Phi}{dN_1}\right) = \Delta N_2 T^2 \frac{\partial}{\partial T}\frac{\mu_1 - \mu_2}{T}.$$

From the expression of the chemical potential, Eq. (4.70), obtained in the previous section, we find however:

$$\frac{\mu_1 - \mu_2}{T} = \frac{\epsilon_1 - \epsilon_2}{T} + \ln(c_1/c_2) - \zeta_1 + \zeta_2,$$

and therefore

$$\Delta H = \Delta N_2(\epsilon_2 - \epsilon_1).$$

We see that the enthalpy variation is positive if $\epsilon_2 > \epsilon_1$, i.e. if the reaction is endothermic. This is the correct result, as ΔH is precisely the heat that should be transferred at constant pressure from the environment to the gas mixture, to keep the temperature constant during the reaction. This amount of heat must necessarily be positive if the reaction absorbs energy. Notice that, in this case, since the number of molecules remains constant during the reaction, in order to maintain the pressure of the system constant, it is sufficient to keep its temperature and its volume constant. In general, if the number of molecules changes, in order to maintain P and T constant, it could be necessary to let the system be free to expand or to compress (see Problem 5).

4.11 Phase Transitions

In this case, by definition, molecules belonging to different phases do no mix. The value of Φ in a region of the system in which there is just one phase of the substance, will be the chemical potential of that particular phase, multiplied by the number of molecules in the region. The total Gibbs free energy will be obtained summing the contributions from the regions in the different phases. In the presence of just two phases (say, liquid and vapor), at fixed P and T:

$$\Phi(P, T, N_l, N_v) = N_l \mu_l(T, P) + N_v \mu_v(T, P), \tag{4.76}$$

and the equilibrium condition will be simply

$$\mu_l(T, P) = \mu_v(T, P).$$

The fact that the chemical potentials are independent of concentrations, has the consequence that Φ will be minimum when all the molecules are in the phase with the smallest μ. The plane PT will be divided in two regions, in which, at equilibrium,

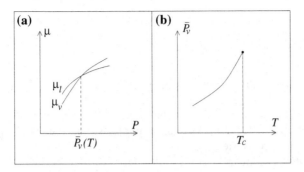

Fig. 4.5 Equilibrium of phases in the plane μP (**a**) and in the plane PT (**b**). Graph **a**: for $P < P_v$, $\mu_v < \mu_l$, and only the vapor phase is possible. Vice versa for $P > P_v$. At equilibrium, we would have therefore $\Phi = N\mu_{l,v}$ to the *right* and to the *left* of P_v, respectively. The discontinuity in the derivatives corresponds to the change of volume $N(v_l - s_v)$ in the change of phase, as illustrated in Fig. 4.6. Graph **b**: the point T_c indicates the critical temperature

only one phase is present (see Fig. 4.5a). The phase transition will occur when the temperature reaches the value $\bar{T}_v(P)$ (or in alternative, when the pressure reaches the value $\bar{P}_v(T)$), such that $\mu_l(T, P) = \mu_l(T, P)$. In the case of a liquid-gas transition, the pressure $\bar{P}_v(T)$ is called the **vapor pressure** at temperature T. (In the case of water, the vapor pressure corresponding to 100 °C is one atmosphere). The fact that the condition of equilibrium between phases, $\mu_l = \mu_v$, is independent of concentration parameters, has the consequence that different phases on the line $P = \bar{P}_v(T)$, can coexist for arbitrary mass ratio N_v/N_l. This corresponds to the situation illustrated in Fig. 4.6: the horizontal portions of the curves at height $\bar{P}_{v,k}$ correspond to different points on the curve $P = P_v(T)$ in Fig. 4.5b; the corresponding isobaric volume changes are associated with the passage from a pure liquid phase (left extreme), to the vapor phase (right extreme).

We thus see that the volume occupied by the system makes a jump when crossing the transition line $P = P_v(T)$ in Fig. 4.5b. A similar discontinuity occurs in the entropy of the system. This change of entropy in the transition liquid-vapor at constant

Fig. 4.6 Isotherms in the presence of a liquid-gas phase transition. The plateau regions identify the corresponding vapor pressure. The *fat dot* indicates the critical point

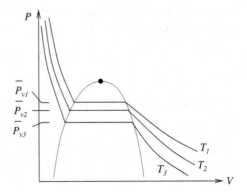

T and P, allows to determine the evaporation enthalpy for the liquid. The procedure is analogous to the one in Sect. 4.10.1 to calculate the reaction enthalpy. Exploiting the fact that at the transition point $\mu_l = \mu_v$, we find:

$$\Delta H = [\mu_l - \mu_v + (s_l - s_v)T]\Delta N_l = (s_v - s_l)T\Delta N_v, \qquad (4.77)$$

where $s_{l,v} = S_{l,v}/N_{l,v}$ are the specific entropy for a molecule in the two phases. As $\Delta N_v = -\Delta N_l > 0$ molecules pass to the vapor phase, an enthalpy $T(s_v - s_l)\Delta N_v$ must be provided to the system. The variation of enthalpy, $\mathscr{L} = T(s_v - s_l)$, defines the **latent heat of evaporation** of per molecule.

Knowing the latent heat of evaporation and the law of state of the substance in the vapor phase, turns out to be enough to determine the temperature dependence of the vapor pressure $\bar{P}_v$. The general idea is that differentiating $\mu_l - \mu_v$ along the transition line $P = P_v(T)$, we must obtain $d(\mu_l - \mu_v) = 0$.

Exploiting the fact that $\mu_{l,v}(T, P)$ are the specific Gibbs free energy for one molecule, in the two phases, we can write, from Eq. (4.76):

$$d\mu_{l,v}(T, P) = \frac{1}{N_{l,v}}d\Phi_{l,v}(T, P) = -s_{l,v}dT + v_{l,v}dP, \qquad (4.78)$$

where $v_{l,v}$ are the specific volumes of the molecules in the two phases.

The condition of moving along the equilibrium line $P = P_v(T)$, corresponding to $d\mu_l = d\mu_v$, will give, using Eq. (4.78):

$$\frac{d\bar{P}_v}{dT} = \frac{s_v - s_l}{v_v - v_l} = \frac{\mathscr{L}}{(v_v - v_l)T}.$$

Now, provided the temperature is sufficiently lower than T_c, we can put $v_v \gg v_l$ and express v_v as function of temperature and pressure, by means of the law of state $PV = NT$:

$$v_v - v_l \simeq v_v = \frac{T}{\bar{P}_v}.$$

Substituting into the expression for $d\bar{P}_v/dT$, we obtain the following relation, called the **equation of Clausius-Clapeyron**:

$$\frac{d\bar{P}_v}{dT} = \frac{\mathscr{L}\bar{P}_v}{T^2}. \qquad (4.79)$$

The Clausius-Clapeyron equation can be solved analytically, and, for small temperature variation, it gives us an exponential dependence of the vapor pressure on the temperature:

$$\bar{P}_v(T) \simeq \bar{P}_v(T_0)\exp\left(\frac{\mathscr{L}(T - T_0)}{T_0^2}\right). \qquad (4.80)$$

Note As illustrated in Figs. 4.5b and 4.6, the phase transition disappears above a certain temperature T_c, called the critical temperature. In the case of water, $T_c \simeq 374\,°C$. At $T = T_c$ the differences $v_v - v_l$ e $s_v - s_l$ vanish, and the difference between phases disappears (more precisely, the transition becomes what is called a second order phase transition; see Sect. 4.11.2). Close to T_c, the approximation $v_v \gg v_l$ is not applicable and the Clausius-Clapeyron equation ceases to be valid.■

4.11.1 Metastable States

The isotherms of a system undergoing a phase change, such as a liquid-gas transition, are characterized by a plateau region in which the system ceases to be homogeneous. This plateau describes a phase equilibrium, in which different phase regions can coexist in the system. One aims to obtaining such isotherms from a microscopic description of the system. A crude example is provided by the van der Waals model, outlined in Appendix A.1.

What one obtains in this way, typically, is isotherm curves describing a homogeneous system, in which a continuous transition takes place between the phases. As illustrated in Fig. 4.7, however, in passing from one phase to the other along the isotherm, the system must cross a region in which the stability condition $(dP/dV)_T < 0$ is violated. To study in detail the stability properties of the homogeneous isotherms, and to determine the vapor pressure $\bar{P}_v$ identifying the plateau, it is convenient to consider a free energy along the line of Eq. (4.62):

$$\hat{\Phi}_{\bar{P}T}(V) = \bar{P}V + F(T, V). \tag{4.81}$$

While the function $\hat{\Phi}_{PT}(S, V)$ in Eq. (4.62) gave the energy that had to be provided to the system to push it to a non-equilibrium state, described by the variables S and

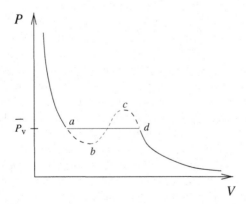

Fig. 4.7 Isotherm at a phase transition. The *continuous line* indicates the stable isotherm, that corresponds to an inhomogeneous state of the system at the transition. *Dashed lines* indicate the unstable (*bd*) and metastable (*ab* and *cd*) intervals, and correspond to a continuous transition, in which the system remains homogeneous

Fig. 4.8 Construction of the
function $\hat{\Phi}_{\bar{P}T}$ for $\bar{P} = \bar{P}_v$.
From Eq. (4.82), it is easy
to check that the difference
between the values of $\hat{\Phi}_{\bar{P}T}$
is given by the difference of
the areas of the two sectors A
and B

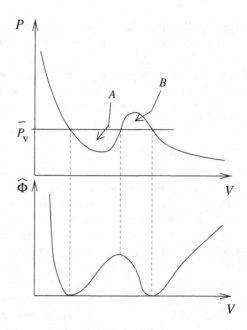

V, $\hat{\Phi}_{PT}(V)$ gives the energy that has to be provided to the system, to push it to a
non-equilibrium state, described by the variable V, in isothermal conditions. The
relation between the free energy $\hat{\Phi}_{PT}(V)$ and the isotherms in Fig. 4.7 is provided
by Eq. (4.43), that gives us

$$\frac{d\hat{\Phi}_{\bar{P}T}(V)}{dV} = \bar{P} - P(V, T). \tag{4.82}$$

As in the case of $\hat{\Phi}_{PT}(S, V)$, stable equilibrium corresponds to a minimum of
$\hat{\Phi}_{\bar{P}T}(V)$. As illustrated in Fig. 4.8, we have two minima, that will be at the same level
only if the area enclosed in the two portions A and B, between the isotherm $P(T, V)$
and the line $P = \bar{P}$, are equal. The condition that the two minima are at the same
level, tells us that the two stable equilibria are equivalent. This corresponds to the
equilibrium of phases, and identifies therefore the vapor pressure $\bar{P}_v$. This is the so-
called **Maxwell construction** of the isotherm for a phase transition: the plateau line
connecting the pure liquid and the pure vapor phases, must lie at a level such that the
two portions of curve $P = P(V, T)$, cut by $P = \bar{P}$, have equal area. If we change the
ambient pressure $\bar{P}$, as illustrated in Fig. 4.9, one of the minima will become lower
than the other. The lower minimum describes the stable state of the system, while the
other is only metastable: the higher the metastable minimum, the smaller the energy
necessary to destabilize it and to push the system to the absolute minimum.

We compare at this point with the isotherm in Fig. 4.7. We see that the sectors
ab and cd corresponds precisely to such metastable situations. In a liquid-gas phase
transition, they would correspond to an overheated liquid and overcooled vapor,

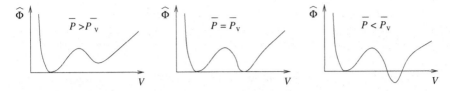

Fig. 4.9 Form of the free energy $\hat{\Phi}$ for three different values of the ambient pressure $\bar{P}$. $\bar{P} > \bar{P}_v$: liquid stable; vapor metastable. $\bar{P} = \bar{P}_v$: both phases are stable (equilibrium of phases). $\bar{P} < \bar{P}_v$: liquid metastable; vapor stable

respectively. A perturbation of a liquid in ab, would cause evaporation and sudden transition to a point at the same level $\bar{P}$ to the right of d. Conversely, perturbation of a gas in a cd, would lead to sudden condensation and transition to a point of the isotherm, at the same level $\bar{P}$, to the left of a. Similarly, if we isothermally compress the gas beyond c, the system would become unstable, and there would be sudden condensation. Likewise, if a liquid expanded isothermally beyond b, there would be sudden evaporation to a point to the right of d.

4.11.2 Critical Phenomena

One of the characteristics of the kind of phase transitions, with which we have been dealing so far, is the presence of a latent heat. The change of state of the system, at constant T and P, is a associated with heat exchange. This results in a discontinuity in the system entropy $S(T, P)$ across the transition line $P = \bar{P}_v(T)$. A similar discontinuity will be present also in the system volume $V(T, P)$. Because of this, both the heat capacity of the system, C_P, and the compressibility, $-V^{-1}(\partial V/\partial P)_T$, will become infinite at the transition line.

Such phase transitions, involving a latent heat, are said to be of **first order**. We have seen that, at the critical point, such discontinuities disappear, together with the distinction between phases. Singular behaviors continue nevertheless to be observed. In fact, both the heat capacity C_P and the compressibility diverge, as the critical point is approached from above, even though no discontinuities in S and V are present. The divergent behavior of $(dV/dP)_T$ is accompanied by fluctuations in the density (i.e. in the specific volume v), with a correlation length λ that becomes infinite at the critical point. Because of this, volume fluctuations do not disappear in the thermodynamic limit. One speaks in this case of a **second order phase transition**.

Broadly speaking, a phase transition is of second order, if no discontinuities in the entropy, or other extensive parameters, such as volume, are present, but divergent behaviors of quantities, with the meaning of a susceptibility, such as C_P and $(dV/dP)_T$, occur, together with divergence of the correlation length for the fluctuations in the system. Right at the critical point, the correlation length becomes infinite, and the correlation function is characterized by a power-law scaling. The fluctuating quantity (more precisely, the fluctuating component of the fluctuating quantity), in the language of critical phenomena, is called an **order parameter**. In the case of the gas-liquid transition, the order parameter is $\phi = v - \langle v \rangle$, with the role of control parameter played by the temperature (or, more typically, its renormalized version $t - 1 - T/T_c$).

The singular behavior of the various quantities near the critical point is parameterized through scaling relations in the form

$$C_V \sim t^{-\alpha}, \quad v_v - v_l \sim (-t)^\beta, \quad -\frac{1}{V}\left(\frac{\partial V}{\partial P}\right)_T \sim t^{-\gamma}, \quad \lambda_\phi \sim t^{-\nu}, \quad \text{etc.} \quad (4.83)$$

The quantities α, β, etc. are called the **critical exponents**. We shall see in Sect. 6.2.2 that they can be expressed one as function of the other, in terms of relations that are independent of the details of the system undergoing the phase transition.

The need for a divergent behavior for $(dV/dP)_T$, near the critical point, can be understood from the behavior of the free energy $\hat{\Phi}_{TP}(\phi)$. The situation is illustrated in Fig. 4.10. For $T > T_c$, the free energy has a single minimum, corresponding to the fact that only one phase exists in the system. From the definition, Eq. (4.81), we see that near this minimum:

$$\hat{\Phi}_{TP}(\phi) \simeq \hat{\Phi}_{TP}(0) + \frac{1}{2}\hat{\Phi}''_{TP}(0)\phi^2 = \hat{\Phi}_{TP}(0) - \frac{N^2}{2}\left(\frac{\partial P}{\partial V}\right)_T \phi^2.$$

At the critical point, we have a bifurcation, in which the original minimum at $T > T_c$, splits into two minima, that progressively separate as $T < T_c$.

Taylor expanding in ϕ, around the transition line (and its prolongation above T_c),

$$\hat{\Phi}_{TP}(\phi) = \hat{\Phi}_{TP}(0) + \frac{1}{2}\hat{\Phi}''_{TP}(0)\phi^2 + \frac{1}{3!}\hat{\Phi}'''_{TP}(0)\phi^3 + \frac{1}{4!}\hat{\Phi}''''_{TP}(0)\phi^4 + \dots,$$

we see that this bifurcation is associated with a change of sign in $\hat{\Phi}''_{TP}(0)$, that becomes zero at $T = T_c$. Thus, divergence of the compressibility for $T \to T_c$, from

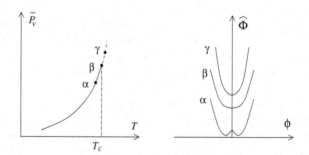

Fig. 4.10 Behavior of the free energy $\hat{\Phi}_{PT}(\phi)$, $\phi = v - \langle v \rangle$, along the transition line. Below T_c, (α), $\hat{\Phi}_{PT}(\phi)$ has two quadratic minima, that correspond to the two phases of the system. The two minima coalesce at the critical point $T = T_c$, (β), in a minimum of quartic order or higher; the distinction between phases disappears. Above T_c, (γ), only one quadratic minimum remains (only one phase present)

above, is associated with the minimum of the free energy becoming quartic at the critical point.

We can see the compressibility, $-V^{-1}(\partial V/\partial P)_T$, as a sort of spring constant for the volume fluctuations. The increase of the fluctuation amplitude for ϕ, near the critical point, is therefore not so surprising. More subtle the origin of the divergence in the correlation length of these fluctuations. We postpone discussion of this issue to Sect. 6.2. For the moment, we observe that arguments analogous to those utilized in Sect. 3.3, allow us to write for the volume fluctuations in a portion of a system containing M molecules:

$$\langle (\Delta V)^2 \rangle = \sum_{ij}^{M} \langle \Delta v_i \Delta v_j \rangle.$$

If the correlation length remained microscopic, one would have

$$\frac{\langle (\Delta V)^2 \rangle}{V^2} \sim M^{-1},$$

and the fluctuations would disappear in the thermodynamic limit. To have fluctuations in the thermodynamic limit, it is necessary that either $\langle (\Delta v)^2 \rangle \to \infty$, as $T \to T_c$, or that the correlation length is itself divergent.

4.12 Thermodynamics of Moist Air

The considerations that we have carried out in Sect. 4.11 could be utilized to determine the conditions of equilibrium in a mixture of water vapor, dry air and water in the liquid phase. This is called a condition of **saturated vapor**, that corresponds to the maximum content of water vapor that air can support at the given temperature.

Let us consider a situation in which the system is at fixed pressure, and utilize as control parameter the humidity content of air $c_v = N_v/(N_v + N_d) \simeq N_v/N_d$, where N_v and N_d are the number of molecules of water vapor and dry air, respectively. Equivalently, we could consider, as control parameter, the partial pressure of water vapor, $P_v = (N_v/N_d)P$, at fixed pressure P. The condition of equilibrium among phases will be again

$$\frac{\partial \Phi}{\partial N_v} = \mu_v - \mu_l = 0,$$

where now $\Phi = N_l \mu_l + N_v \mu_v + N_d \mu_d$. Obviously, the dry component $N_d \mu_d$ does not contribute in the transition.

In analogy with what was done in the calculation of the vapor pressure, we seek a differential equation for the vapor concentration at equilibrium, $c_v^{sat}(T)$, by differentiating the condition $\mu_l - \mu_v = 0$ with respect to T e N_v at fixed P. In alternative, exploiting the relation $P_v = c_v P$, we can differentiate with respect to T e P_v, again at fixed P. This brings us back to Eq. (4.78):

$$0 = \frac{\partial(\mu_v - \mu_l)}{\partial N_v} dN_v + \frac{\partial(\mu_v - \mu_l)}{\partial T} dT = \frac{\partial \mu_v}{\partial N_v} dN_v + \frac{\mathscr{L} dT}{T}. \tag{4.84}$$

(We have hypothesized that the liquid phase is pure, so that μ_l does not contain terms dependent from N_d; compare with Eq. (4.70)). Exploiting Eq. (4.72), we find however:

$$\frac{\partial \mu_v(T, P, c_v)}{\partial N_v} dN_v = \frac{\partial \mu_v(T, P_v)}{\partial P_v} dP_v = v_v dP,$$

where, from Eq. (4.71), $v_v = T/P_v$. Substituting into Eq. (4.84), we obtain the following variation over the Clausius-Clapeyron equation:

$$\frac{dP_v^{sat}}{dT} = \frac{\mathscr{L} P_v^{sat}}{T^2}.$$

In other words, we have saturation when the partial pressure of water vapor in air, P_v, becomes equal to the vapor pressure of water at the given temperature, $\bar{P}_v(T)$. From here, we obtain the humidity content at saturation

$$\frac{dc_v^{sat}}{dT} = \frac{\mathscr{L} c_v^{sat}}{T^2}, \tag{4.85}$$

that, combined with Eq. (4.80), tells us that humidity at saturation grows exponentially with temperature. The value of the temperature, at which humidity at saturation becomes equal to the humidity content of the air, is called **dew point**.

4.13 Problems

Problem 1 A volume of air in atmospheric conditions is compressed adiabatically from V to $V/2$. Discuss the conditions under which the process can be considered as quasistatic.

Solution In quasi-static conditions, the pressure must satisfy instantaneously the adiabatic law, Eq. (4.8). This means that the fluid motions generated by the compression, produce negligible modifications in the pressure. A sufficient condition for this, is that the process is slow enough for viscosity to damp any velocity perturbation. This requires that the Reynolds number, obtained putting together the space and time scales of the process, and the air viscosity, must be small:

$$Re = \frac{V^{2/3}}{t\nu} \gg 1.$$

where t is the time required by the process, and $V^{1/3}$ is the characteristic length. We obtain the condition on the compression time

$$t \gg \frac{V^{2/3}}{\nu},$$

This condition, however, turns out to be too strong, as internal motion, even when present, may be not energetic enough to produce relevant changes in the pressure. Let us consider therefore the opposite limit in which viscosity does not play a role. We can obtain the pressure perturbation ΔP balancing inertia and pressure in the Navier-Stokes equation

$$\frac{\Delta P}{V^{1/3}} \sim \varrho \frac{V^{1/3}}{t^2},$$

where the velocity scale of the process is provided by $u = V^{1/3}/t$. This gives us, from the condition for a quasistatic process, $\Delta P \ll P$:

$$\frac{V^{2/3}}{t^2} \ll \frac{P}{\varrho} \sim c_s^2,$$

where c_s is the speed of sound of air at atmospheric conditions. The quasistatic condition is therefore that the Mach number of the process $\mathrm{Ma} = u/c_s$, is small.

Problem 2 A mass M of an ideal monoatomic gas is heated, at constant pressure, from temperature T to temperature $T + \Delta T$. Calculate the variation of the entropy and of the volume of the gas.

Problem 3 We have two amounts of an identical ideal monoatomic gas, separated by a wall. The temperature of the two gases is the same, but the pressure is different. Suppose that the wall is lifted. Calculate the change of entropy in the system.

Problem 4 An ideal gas is contained in a cylinder with thermally conducting walls, in contact with a thermal bath at temperature T. Inside the cylinder there is a piston. Let $V_{a,b}$ and $P_{a,b}$ be the volume and the pressure of the gas on the two sides of the piston. Calculate the change in the Helmholtz free energy in the system, after the cylinder is left free to move.

Problem 5 A mixture of ideal gases, A, B and C, is characterized initially by values $c_A = N_A/N$, $c_B = N_B/N$ and $c_C = N_C/N$, of the concentration. Suppose that the chemical constants, ζ_i, and the chemical energies of the molecules, ϵ_i, are independent of pressure and temperature. The three gases can undergo the chemical reaction $A + 2B \leftrightarrow C$. Assume that the reaction takes place at constant T and P, and that initially $c_B = 2c_A$ and $c_C = 0$. Calculate the concentrations $\bar{c}_i$ at equilibrium and discuss their dependence on P, T and on the constants ϵ_i and ζ_i.

Solution The chemical equilibrium condition provided by Eq. (4.74), given the stoichiometric ratio of the reaction in the problem, is

$$\mu_A + 2\mu_B - \mu_C = 0.$$

We notice at once that the stoichiometric ratio in the reaction, guarantees that the concentration ratio $c_B/c_A = 2$ remains constant. Exploiting $c_A + c_B + c_C = 1$, we can write

$$c_A = \frac{1}{3}(1 - c_C) \quad \text{and} \quad c_B = \frac{2}{3}(1 - c_C).$$

Using Eq. (4.70), the chemical equilibrium condition becomes

$$5T(1 - \ln T) + 2T \ln P + \epsilon_A + 2\epsilon_B - \epsilon_C - T(\zeta_A + 2\zeta_B - \zeta_C) = -T \ln \frac{c_A c_B^2}{c_C},$$

that can be rewritten in the form

$$\frac{c_C}{(1 - c_C)^3} = \frac{4e^5}{9} \frac{P^2}{T^5} \exp\left(\zeta_C - \zeta_A - 2\zeta_B - \frac{\epsilon_C - \epsilon_A - 2\epsilon_B}{T}\right).$$

Pressure contributes to chemical equilibrium, with higher pressure corresponding (for this kind of reaction) to a higher concentration of reaction products.

4.14 Further Reading

Most of this section is a rephrasing of material contained in classical books such as:

- E. Fermi, *Thermodynamics* (Dover, 1936)
- L.D. Landau, E.M. Lifsits, *Statistical Physics* (Pergamon Press, 1969)
- H.B. Callen, *Thermodynamics and an Introduction to Thermostatistics*, 2nd edn. (Wiley, 1985)

Additional reference books:

- C. Guthman, B. Roulet, B Diu, D. Lederer, *Physique Statistique* (Hermann, 1997)
- R.H. Swendsen, *An Introduction to Statistical Mechanics and Thermodynamics* (Oxford, 2012)

Negative temperature states and negative C_V systems are discussed e.g. in:

- Y. Levin, R. Pakter, F.B. Rizzato, T.N. Teles, F.P.C. Benetti, Non-equilibrium statistical mechanics of systems with long-range interactions. Phys. Rep. **535**, 1 (2014)

Appendix

A.1 Negative Temperature States

The exotic nature of negative temperature states, does not prevent them from existing in nature. Typically, they describe non-equilibrium conditions of physical systems.

Negative temperatures become possible e.g. in systems composed of microscopic constituents, that have a finite number of discrete energy levels $\{\epsilon_1, \epsilon_2, \ldots \epsilon_M\}$. If the microscopic constituents do not interact, the energy of the system will be the sum of the energies of the individual constituents:

$$E = \sum_{i=1}^{N} \epsilon_{k(i)},$$

where $k(i)$ is the state of the ith component. Thus, by construction, $E_{min} < E < E_{max}$ with $E_{min} = N\epsilon_1$ and $E_{max} = N\epsilon_M$.

The entropy of the system, at the given energy E, will be the Shannon entropy of the joint probability distribution $P(\epsilon_{k(i)}, i = 1, \ldots N\}|E)$. It is easy to be convinced that $S(E_{min}) = S(E_{max}) = 0$, as $E = E_{min}$, requires that all microscopic components have $\epsilon = \epsilon_1$, while $E = E_{max}$ requires that all microscopic components have $\epsilon = \epsilon_M$ (in both cases, we know exactly the state of all components). Since $S(E) \geq 0$, there will be some energy intervals in which $T^{-1} = S'(E) < 0$.

A condition that could be interpreted as a negative temperature state is the population inversion that is realized in the active medium of a laser, by optical pumping by an external light source.

Systems with a negative heat capacity exist as well in nature. A classical example is that of a self-gravitating body. In this case, the sum of the kinetic and potential energies of all the molecules, K and U are related through the virial theorem:

$$2K = -U.$$

One has therefore for the total internal energy of the body:

$$E = K + U = -K.$$

Temperature continues to be related to thermal motion through a relation $T \propto K$ (see Sect. 5.6). Hence we have for the heat capacity of the body

$$C_V = \frac{\partial E}{\partial T} \propto \frac{\partial E}{\partial K} < 0.$$

The body becomes hotter by loosing its energy.

This phenomenon has deep cosmological implication, as it is taken as possible explanation of the fact that the universe, at nearly 14 billion years of age, is still in a state very far from thermodynamic equilibrium. The negative heat capacity of self-gravitating objects, suggests that, at scales in which gravitation is the dominant interaction, systems do not evolve to a spatially homogeneous state. The formation of stars and of agglomerates of stars, could be a manifestation of this tendency towards an heterogeneous state.

A.2 The Van der Waals Model

The simplest model of a real gas is due to J.D. van der Waals. Its main success was the ability to qualitatively describe the dynamics of the gas-liquid phase transition. The model is based on just two ingredients

- An attractive binary long-range interaction, resulting in a decrease in the pressure. This effect is taken to be independent of temperature, as it is generated by packing of the molecules, independent of thermal motion. Being the result of a binary interaction, it will be proportional to the square of the density of the density, i.e. to the inverse square of the molecular specific volume $v = V/N$.
- A hard core repulsive short range interaction: molecules cannot get closer than an effective diameter d. The result is that the actual volume available to a molecule of a real gas, will not be the total volume V occupied by the gas, rather, the subtracted amount $V_{eff} = V - Nb$, where $b \sim d^3$ is the effective molecular volume. Thus, the thermal motion contribution to pressure, will not be weighed by $n = 1/v$, rather, by $(v - b)^{-1}$.

The two ingredients are put together in a law of state in the form

$$P = \frac{T}{v - b} - \frac{a}{v^2},\tag{4.86}$$

where $v = V/N$ is the specific volume of the molecules. We see that the law of state of ideal gases is recovered in the limit $v \to \infty$. In the other limit, there is a competition between the contribution by the repulsive and attractive components of the interaction, with the first becoming dominant as $v \to b$.

The behavior of the isotherms described by Eq. (4.86) is illustrated in Fig. 4.11. We see that for T sufficiently large, the long-range part of the interaction (the term proportional to a) does not contribute, and the profile is that of an ideal gas isotherm horizontally shifted by b. For T small, on the other hand, there is a range of values for v, in which the attractive term produces a local minimum for the pressure. The isotherm corresponding to formation of the local minimum, identifies the critical temperature of the system, that we can verify being determined by the values of the parameters

$$T_c = \frac{8a}{27b}; \qquad P_c = \frac{a}{27b^2}; \qquad v_c = 3b.$$

Fig. 4.11 Sketch of van der
Waals Isotherms at temperatures $T_1 > T_c > T_2$

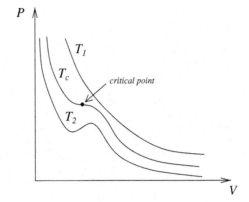

Expressing temperature, pressure and specific volume as function of their critical
values, $T = \hat{T}T_c$, $P = \hat{P}P_c$ and $v = \hat{v}v_c$, Eq. (4.86) can be written in a universal
form, independent of the parameters of the gas:

$$\hat{P} = \frac{\hat{T}}{\hat{v} - 1/3} - \frac{3}{\hat{v}^2}, \tag{4.87}$$

The fact that the van der Waals law can be expressed in such a universal form,
independent of the gas, is called the **law of correspondent states**.

Chapter 5
Introduction to Equilibrium Statistical Mechanics

5.1 The Microcanonical Distribution

The only microscopic description of thermodynamic systems that we have seen so far is the one provided by kinetic theory. That of kinetic theory, is an approach applicable in principle to a great variety of systems, both in equilibrium and in non-equilibrium conditions, but only at the price of closure approximations on the many-particle statistics, that are often difficult to control.

The statistical mechanics approach, that we are going to describe now, avoids the need of closure approximations on the microscopic statistics, by restricting the analysis to Hamiltonian systems in thermodynamic equilibrium conditions. The basic idea is that the thermodynamic equilibrium state of a macroscopic system must correspond to a regime of stationary statistics at microscopic scales. The comparatively simpler structure of Hamiltonian phase space allows to determine the stationary distribution $\rho(\Gamma)$ in closed form,[1] without the need of closure approximations. Once the stationary distribution $\rho(\Gamma)$ is known, the equilibrium values of the thermodynamic variables $\mathbf{m}$ can be obtained as averages in Γ-space in explicit form:

$$\bar{\mathbf{m}} \simeq \langle \mathbf{m} \rangle = \int d\Gamma \, \tilde{\mathbf{m}}(\Gamma) \rho(\Gamma). \tag{5.1}$$

The physical interpretation of Eq. (5.1) is analogous to that of the corresponding formula in kinetic theory, Eq. (3.1). We can understand Eq. (5.1) as an ensemble average over a collection of identical systems, each prepared in the identical macroscopic condition. Alternatively, we could imagine to perform a time average on a single system, assuming that some kind of ergodic property for the variable $\mathbf{m}$ is satisfied:

$$\bar{\mathbf{m}} = \lim_{T \to \infty} \frac{1}{T} \int_0^T dt \, \tilde{\mathbf{m}}(\Gamma(t)), \quad \forall \Gamma(0), \quad \mathscr{H}(\Gamma(0)) = E, \tag{5.2}$$

---

[1] For lighter notations, we drop from now on subscript N on the distributions in Γ-space.

© Springer International Publishing Switzerland 2015
P. Olla, *An Introduction to Thermodynamics and Statistical Physics*,
UNITEXT for Physics, DOI 10.1007/978-3-319-06188-7_5

where $\mathscr{H}$ is the Hamiltonian of the system. We see that ergodicity describes here simply uniqueness of the thermal equilibrium state associated with energy E. In this picture, relaxation to thermodynamic equilibrium corresponds to a phase point Γ that moves from a low-probability non-equilibrium region of Γ-space, to a high-probability equilibrium one.

Unfortunately, giving physical substance to the averages in Eqs. (5.1) and (5.2), is not simple. The size of the statistical sample that would be necessary to characterize a PDF in an N-dimensional Γ-space grows exponentially with N. Thus, the only way the averages in Eqs. (5.1) and (5.2) can make any sense, is if $\tilde{\mathbf{m}}(\Gamma) \simeq \bar{\mathbf{m}}$ in most of the part of Γ-space in which the probability mass of $\rho(\Gamma)$ is concentrated. Once this condition is satisfied, the exact form of the PDF $\rho(\Gamma)$ is essentially irrelevant. Actually, it is not even necessary that the stationary distribution $\rho(\Gamma)$ is unique, or that it is the equilibrium distribution for the system.

In any case, it turns out that the stationary distribution $\rho(\Gamma)$ can be determined explicitly, provided the microscopic dynamics of the isolated system is Hamiltonian. We can prove in fact that any PDF in the form

$$\rho(\Gamma) = f(\mathscr{H}(\Gamma)), \tag{5.3}$$

with f a generic (non-negative, normalized) function, will be stationary.

Let us verify this result.

Consider an isolated Hamiltonian system with N degrees of freedom. This system will be described by $2N$ canonical coordinates $\Gamma \equiv (q_i, p_i; i = 1, \ldots N)$ (in the case of a monoatomic gas, for the ith molecule we will have three coordinates $\{q_{i1}, q_{i2}, q_{i3}\} \equiv \{x_{i1}, x_{i2}, x_{i3}\}$ and three conjugated moments $\{p_{i1}, p_{i2}, p_{i3}\} \equiv m\{v_{i1}, v_{i2}, v_{i3}\}$, i.e. $3N$ degrees of freedom for $6N$ canonical coordinates).

The evolution of the PDF $\rho(\Gamma, t)$ is determined by the continuity equation for the Hamiltonian flow

$$\dot{\Gamma} = (\dot{q}_i, \dot{p}_i, i = 1, \ldots) = \left(\frac{\partial \mathscr{H}}{\partial p_i}, -\frac{\partial \mathscr{H}}{\partial q_i}, i = 1, \ldots N \right). \tag{5.4}$$

We can visualize this flow in Γ space, with the aid of the concept of ensemble, as a flow of phase points moving along trajectories determined by the Hamilton equations for the system. The corresponding continuity equation is called the **Liouville equation**:

$$\frac{\partial \rho}{\partial t} + \nabla_\Gamma \cdot (\rho \dot{\Gamma}) = 0. \tag{5.5}$$

Here,

$$\nabla_\Gamma = \left(\frac{\partial}{\partial q_i}, \frac{\partial}{\partial p_i}, i = 1, \ldots N \right) \tag{5.6}$$

indicates the $2N$ dimensional gradient operator in Γ-space.

At stationary state, the Liouville equation can be written in the form

$$\dot{\Gamma} \cdot \nabla_\Gamma \rho = -\rho \nabla_\Gamma \cdot \dot{\Gamma},$$

that will admit a uniform solution in the Γ-space if the flow $\dot{\Gamma}$ is incompressible: $\nabla_\Gamma \cdot \dot{\Gamma} = 0$.

This is in fact the case: the Hamiltonian flow is incompressible, which is the content of the so-called **Liouville theorem**. We can easily verify the result. From Eqs. (5.4) and (5.6), we find

$$\nabla_\Gamma \cdot \dot{\Gamma} = \sum_i \left(\frac{\partial \dot{q}_i}{\partial q_i} + \frac{\partial \dot{p}_i}{\partial p_i} \right) = \sum_i \left(\frac{\partial}{\partial q_i} \frac{\partial \mathscr{H}}{\partial p_i} - \frac{\partial}{\partial p_i} \frac{\partial \mathscr{H}}{\partial q_i} \right) = 0,$$

and therefore the uniform solution in Γ space is also stationary. Moreover, since the Hamiltonian flow takes place on the $2N - 1$ dimensional energy surfaces $\mathscr{H}(\Gamma) = E$, any solution ρ that is locally constant in energy shells $\Delta\Gamma(E)$, corresponding to energy intervals $[E, E + \Delta E]$, will be stationary as well (by energy conservation, the Hamiltonian flow remains confined in the shells). Taking the $\Delta E \to 0$ limit, we obtain the result that the generic stationary solution of the Liouville equation is in the form given in Eq. (5.3).

Now, an isolated system will always have a well defined energy. It has therefore some sense to consider a PDF that is non-zero only in a particular energy shell

$$\rho^M(\Gamma) = \begin{cases} (\Delta\Gamma(E))^{-1}, & \mathscr{H}(\Gamma) \in [E, E + \Delta E], \\ 0 & \text{otherwise}, \end{cases} \tag{5.7}$$

where we have used the same symbol $\Delta\Gamma(E)$ to indicate both the Γ-space region and its volume. The parameter ΔE identifies here an arbitrary coarse-graining energy scale (the precision with which we want to determine the energy of the macroscopic system). The PDF introduced in Eq. (5.7) is called **microcanonical** and will play a fundamental role in the derivation of statistical mechanics in the pages that follow.

5.1.1 The Ergodic Problem

A conceptual difficulty of the statistical mechanics approach is the absence of a direct physical meaning for the PDF $\rho(\Gamma)$. The situation is different from that of kinetic theory, in which a stationary distribution (actually, the equilibrium distribution) could be obtained explicitly

$$\rho(\Gamma) \simeq \prod_{i=1}^N \rho_1(\mathbf{y}_i),$$

and had a direct physical interpretation thanks to the macroscopic nature of the one-particle distribution $f_1 = N\rho_1$. A possibility to give a physical meaning to the PDF $\rho(\Gamma)$, could be see it as the limit of a macroscopic PDF $\rho(\mathbf{m})$, in which the number of molecules in the subsystems to which the components of the vector $\mathbf{m}$ refer, is sent to zero. Unfortunately, the existence of such a limit is not guaranteed at all. Neither is uniqueness, which is connected with satisfaction of ergodicity of the flow in Γ-space.

Ergodicity of the flow in Γ-space is a problem that has attracted the attention of scientists over the years. We recall from Sect. 2.5.1 that a condition for ergodicity of a stochastic process $x(t)$, was indecomposability of its configuration space: there must be no regions of the domain of x that are inaccessible from other regions of the same domain.

In the case of a deterministic Hamiltonian system, this same property holds, which is the content of the Birkhoff theorem. Central to the result, the so called Poincaré recurrence theorem: if phase space is indecomposable, almost all trajectories will return, after a sufficiently long time, arbitrarily close to their starting point. If the phase-space is indecomposable, the average of any function $f(\Gamma)$ can then be evaluated as a limit of a time average:

$$\langle f \rangle = \lim_{T \to \infty} \frac{1}{T} \int_0^T dt \, f(\Gamma(t)), \qquad \forall \Gamma(0), \quad \mathcal{H}(\Gamma(0)) = E. \qquad (5.8)$$

In a certain sense, we can say that ergodicity in a Hamiltonian system is violated already by energy conservation, as a phase point cannot move into regions of phase space with different energy. The answer to this was of course the decision to restrict to constant energy regions of Γ-space, which lead to the introduction of the microcanonical distribution.

The ergodic question regards precisely what would happen in case of existence of other conserved quantities, beside energy (and possibly the total linear and angular momentum of the system). An answer to this question was provided by Khinchin [A.I. Khinchin, Mathematical foundations of statistical mechanics (Dover, 1949)]. It can be summarized (perhaps in slightly tautologic way) in the following two observations:

• If the conserved quantity has a macroscopic meaning, the microcanonical ensemble should be generalized to take into account conservation of this quantity. Thus ρ^M will be uniform in shells at constant values of the energy and of the additional conserved quantity.

• If the quantity does not have a macroscopic meaning, the choice of the distribution of its values can be carried out in arbitrary manner, and the result of the averages of macroscopic quantities, by definition, must not change. In particular, the choice in which the values of this quantity are distributed in such a way to reproduce the original energy-based microcanonical distribution, is perfectly acceptable.

In other words, the ergodic property for the generic macroscopic quantity **m** must continue to be satisfied even in the presence of conserved microscopic quantities. If $\mathscr{I}$ indicates the value of the microscopic conserved quantity at the initial time, we will thus have

$$\lim_{T \to \infty} \langle \mathbf{m} | \mathscr{I} \rangle_T = \langle \mathbf{m} | \mathscr{I} \rangle = \langle \mathbf{m} \rangle_M, \tag{5.9}$$

where the last average is calculated with respect to the energy-based microcanonical distribution. Conversely, dependence of an average $\langle f | \mathscr{I} \rangle$ on the microscopic condition $\mathscr{I}$ (with consequent break-up of the ergodic condition, Eq. (5.8)), will signal the fact that f must be itself a microscopic quantity.

All these considerations confirm the observations already made at the beginning of Sect. 5.1:

- The exact form of the stationary PDF in Γ-space may not be that important. What is important is that the macroscopic properties of the system be constant in much of the dominion $\Delta\Gamma(E)$.
- The main advantage of the hypothesis $\rho(\Gamma) = \rho^M(\Gamma)$ is perhaps just ease in the calculations.

5.1.2 Statistical Equilibrium and Mixing

We have seen that the microcanonical PDF $\rho^M(\Gamma)$ is stationary. We have not discussed yet, however, whether it is also the equilibrium distribution for the microscopic dynamics.

This is a substantial question: the microcanonical distribution realizes the maximum of the Shannon entropy. The bulk of this distribution corresponds to thermodynamic equilibrium. Thus, even though statistical mechanics is a strictly equilibrium theory, it would be nice that there were a correspondence between thermodynamic equilibrium at the macroscale, and statistical equilibrium at the microscale. This also to make contact with the kinetic theory approach, in which relaxation to thermodynamic equilibrium (with all the limitations of a closure based approach), was looked at precisely in this fashion.

We have to check whether it is true that any initial condition ρ will evolve, after a sufficiently long time, into the microcanonical distribution ρ^M. Notice that this process could be interpreted as a loss of memory of the initial microscopic state of the system (see discussion in Sect. 4.2). The condition for existence of an equilibrium distribution, discussed in Sect. 2.5, in the case of a stochastic process, must however be adapted to the present situation of a deterministic Hamiltonian system.

We see that coarse graining of the distribution, becomes an essential ingredient of the procedure. An example illustrates the kind of problem we are faced with. Consider an initial distrubution $\rho(\Gamma, 0)$ that is different from zero and constant, only in a region of Γ-space $\Delta\Gamma_0 \subset \Delta\Gamma(E)$. The Shannon entropy of this distribution will be, from Eq. (3.21):

$$S(0) = \ln(\Delta\Gamma_0/(N!\delta\Gamma)).$$

As time passes, the points in $\Delta\Gamma_0$ will be transported by the Hamiltonian flow until, at time t, they will occupy a region $\Delta\Gamma_t$. Since the Hamiltonian flow is incompressible, however, the volumes of $\Delta\Gamma_0$ and $\Delta\Gamma_t$ will be the same, as will be identical the value of the distributions in the respective supports:

$$\rho(\Gamma \in \Delta\Gamma_0, 0) = \rho(\Gamma \in \Delta\Gamma_t, t) = (\Delta\Gamma_0)^{-1}.$$

The only possibility for the phase points to fill the domain $\Delta\Gamma(E)$, is that, in some way, the flow in Γ-space has a mixing character, in the sense described in Fig. 3.10.3. This means that, in any small volume of Γ space, after a sufficiently long time, it should be possible to find points initially in $\Delta\Gamma_{\rho_0}$. The mechanism leading to this, is stretching and distortion of $\Delta\Gamma_0$, until it becomes a very long, thin and convoluted ($6N$-dimensional) filament. If the flow is also ergodic, the filaments will be distributed uniformly in $\Delta\Gamma(E)$.

It is clear however, that, in order to say that $\rho \to \rho^M$, our PDF's must be defined at a fixed coarse-grained scale, that blurs filaments thinner than this scale, into a uniform indistinct object, that fills most of $\Delta\Gamma(E)$.

Note We are faced again with the fact that, in order to derive the thermodynamic properties of a macroscopic system from its microscopic dynamics, some kind of coarse graining is necessary. Coarse graining has a deep effect on the way we understand irreversibility. Without coarse graining, irreversibility would be simply a statement on trajectories in Γ-space. Since the underlying Hamilton equations are time-symmetric, the source of asymmetry would lie in the choice of initial condition for the trajectory. Poincaré recurrence guarantees however that the isolated system will return, after a sufficient long time (indeed much longer than the life of the universe), arbitrarily close to the initial non-equilibrium condition. Thus, considered in a sufficiently long time interval, the trajectory $\Gamma(t)$ would loose its irreversible character.

Coarse-graining translates irreversibility, from a statement over trajectories, to one over PDF's. In this case, the evolution of $\rho \to \rho^M$ becomes fully irreversible, even though the single trajectories alternate between equilibrium and non-equilibrium states.

We point out that a similar operation was carried out implicitly also in kinetic theory, modeling collisions as a Markov process, and then disregarding the correlations among the particles participating in the process (mean field approximation). In kinetic theory, information was destroyed coarse-graining the collision process. In statistical mechanics, the same result is obtained disregarding the fine grained structure of ρ in Γ-space.■

5.2 Entropy in Equilibrium Statistical Mechanics

The discussion that we have carried out so far, suggests that the only way to give a physical meaning to a PDF in Γ-space is through calculation of averages of macroscopic quantities. In this picture, a central role is played by the concept of entropy. To this aim, we shall introduce a new kind of entropy, expressed as function of macroscopic quantities, that, from one side, will allow to establish a connection between microscopic world (Shannon entropy) and macroscopic world (thermodynamic entropy), from the other, it will allow to better clarify the meaning of the microcanonical distribution as an object that describes situation of both thermodynamic equilibrium and non-equilibrium.

Let us start by giving some terminology.

We indicate with $\mathbf{m} = \{m_a, a = 1, \ldots, M\}$ the vector of the parameters with which we want to describe the macroscopic state (**macrostate**) of the system. We call the ambient space of the vectors $\mathbf{m}$, the μ-space of the system. The components m_a are supposed known with precision Δm_a, so that $\mathbf{m}_a$ is actually a discrete variable. As the microscopic state (**microstate**) of the system is identified by a set of cells of Γ-space of total size $N!\delta\Gamma \sim N!\hbar^{3N}$,[2] the macrostate will identify a cell of size Δm of μ-space. The choice of the parameters that determine the macrostate is arbitrary: $\mathbf{m}$ could be the vector, whose only two components are the number of molecules and the energy in a small volume of the system. Alternatively, it could be the vector with components $f(\mathbf{x}_a, \mathbf{v}_a)(\Delta x)^3(\Delta v)^3$, each of which is the numbers of molecules in the cell of the one-molecule phase space $(\Delta x)^3(\Delta v)^3$, around $(\mathbf{x}_a, \mathbf{v}_a)$. Many other choices are of course possible. In any case, it is clear that, in order to have a macroscopic meaning, the parameters m_a must refer to macroscopic parts of the system, that contain a number of molecules sufficient to guarantee satisfaction of the thermodynamic limit.

Let us indicate with $\Delta\Gamma(\mathbf{m})$ the part of Γ-space corresponding to the macrostate $\mathbf{m}$. For simplicity, we limit the analysis to the case in which $\Delta\Gamma(\mathbf{m}) \subset \Delta\Gamma(E)$, so that $\cup_{\mathbf{m}}\Delta\Gamma(\mathbf{m}) = \Delta\Gamma(E)$. From Eqs. (5.1) and (5.7), the probability to observe a macrostate $\mathbf{m}$, is just the volume fraction of the energy shell $\Delta\Gamma(E)$ corresponding to that macrostate

$$P(\mathbf{m}) = \frac{\Delta\Gamma(\mathbf{m})}{\Delta\Gamma(E)}. \tag{5.10}$$

The restriction to the domain $\Delta\Gamma(\mathbf{m})$ of the microcanonical PDF is $\rho^M(\Gamma|\mathbf{m}) = 1/\Delta\Gamma(\mathbf{m})$, and we can write, for $\Gamma \in \Delta\Gamma(\mathbf{m})$: $\rho^M(\Gamma) = P(\mathbf{m})\rho^M(\Gamma|\mathbf{m})$. From here, we see that the Shannon entropy associated with the microcanonical distribution, for a system of N identical particles, can be written in the form of Eq. (3.21):

---

[2] To fix the ideas we continue to restrict the analysis to the dilute gas case, although the derivation that follows does not depend on the internal structure of the microstates (see discussion in Sect. 5.3.1).

$$S = - \int_{\Delta\Gamma(E)} d\Gamma \, \rho^M(\Gamma) \ln[\rho^M(\Gamma)N!\delta\Gamma] = \ln[\Delta\Gamma(E)/(N!\delta\Gamma)]$$

$$= \sum_{\mathbf{m}} P(\mathbf{m}) S(\mathbf{m}) - \sum_{\mathbf{m}} P(\mathbf{m}) \ln P(\mathbf{m}), \qquad (5.11)$$

where—$\sum_{\mathbf{m}} P(\mathbf{m}) \ln P(\mathbf{m})$ is the Shannon entropy of the distribution of the macroscopic variable $\mathbf{m}$, and

$$S(\mathbf{m}) = - \int_{\Delta\Gamma(\mathbf{m})} d\Gamma \, \rho^M(\Gamma|\mathbf{m}) \ln[\rho^M(\Gamma|\mathbf{m})N!\delta\Gamma] = \ln[\Delta\Gamma(\mathbf{m})/(N!\delta\Gamma)] \quad (5.12)$$

is the Shannon entropy obtained considering $\Delta\Gamma(\mathbf{m})$ as our sample space. We shall call this quantity the **Boltzmann entropy** of the macrostate $\mathbf{m}$. The quantity $\Delta\hat{\Gamma}(\mathbf{m}) = \Delta\Gamma(\mathbf{m})/(N!\delta\Gamma)$ is the number of microstates with which the macrostate $\mathbf{m}$ is realized. Notice that Eq. (5.12) allows us to write the PDF in Γ-space as a function of entropy

$$\rho^M(\Gamma|\mathbf{m}) \sim \exp(-S(\mathbf{m})), \qquad (5.13)$$

and the microcanonical PDF $\rho^M(\Gamma)$ will be obtained putting $\mathbf{m} = E$.

We have already met an object similar to $S(\mathbf{m})$ in Sect. 3.4: the kinetic entropy, Eq. (3.23), and, in particular, its instantaneous counterpart, Eq. (3.25). Also in that case, entropy described the information content of a macrostate, starting from the properties of a PDF in Γ-space. While in kinetic theory, the macrostate $\mathbf{m}$ was the result of the choice of a particular non-equilibrium PDF $\rho(\Gamma, t) \simeq \prod_k \rho_1(\mathbf{x}_k, \mathbf{v}_k; t)$, in statistical mechanics, it must be seen as a fluctuation state in a statistical equilibrium situation, described by the microcanonical distribution.

We point out that, while $\rho(\Gamma, t)$ could describe a non-equilibrium state also with respect to variables $\mathbf{m}'$ not accounted for in $\mathbf{m}$, $\rho^M(\Gamma|\mathbf{m})$, being the restriction of ρ^M to $\Delta\Gamma(\mathbf{m})$, will by construction be dominated by the equilibrium configurations for $\mathbf{m}'$ (provided $\mathbf{m}$ and $\mathbf{m}'$ are independent). Another difference lies in the fluctuations of $\mathbf{m}$ (in the case of $\rho^M(\Gamma|\mathbf{m})$, they are absent by definition).

The two distributions $\rho(\Gamma, t)$ and $\rho(\Gamma|\mathbf{m})$ will become macroscopically equivalent, by definition, in the absence of differences in the way any additional macrostate $\mathbf{m}'$ is distributed. Thus, we expect that $S(\mathbf{m})$ and the kinetic entropy corresponding to the same macrostate should coincide. We shall verify explicitly this point in Sect. 5.3, in the case $m_a = f_1(\mathbf{x}_a, \mathbf{v}_a; t)(\Delta x)^3(\Delta v)^3$.

We call attention to the following important property of both the Shannon entropy, Eq. (5.11), and the Boltzmann entropy Eq. (5.12), in the case of a system described by the microcanonical distribution:

• The entropy corresponding to the macrostate $\mathbf{m}$ is equal to the logarithm of the number of microstates

$$\Delta\hat{\Gamma}(\mathbf{m}) = \Delta\Gamma(\mathbf{m})/(N!\delta\Gamma),$$

with which the macrostate is realized.

Now, if the thermodynamic limit is satisfied by the parts of the system identified by the parameters m_a, and if the discretization constant Δm_a is not too small, only one macrostate will be observed: the most probable one, corresponding to thermodynamic equilibrium $\bar{\mathbf{m}} \simeq \langle \mathbf{m} \rangle$.[3] In other words, we have $P(\bar{\mathbf{m}}) \simeq 1$, so that, combining with Eqs. (5.11) and (5.12):

$$S \simeq \bar{S}(\mathbf{m}) \simeq S(\bar{\mathbf{m}}) \simeq \ln[\Delta \hat{\Gamma}(\bar{\mathbf{m}})], \tag{5.14}$$

where the Boltzmann entropy $S(\bar{\mathbf{m}})$ corresponding to the equilibrium macrostate $\bar{\mathbf{m}}$ is sometimes called **Gibbs entropy**. We notice that the Boltzmann entropy $S(\mathbf{m})$ is a fluctuating quantity, while the Shannon entropy is a constant, that is approximated by the equilibrium value $S(\bar{\mathbf{m}})$. Again, the microcanonical PDF can be recovered substituting into Eq. (5.13), the Gibbs entropy $S(\bar{\mathbf{m}})$. It is also important to stress that, if the condition $P(\bar{\mathbf{m}}) \simeq 1$, and then also Eq. (5.14), are satisfied, the Gibbs entropy will become independent of the macroscopic variables with respect to which it was defined.

Summarizing, we have found the following important results:

- Thermodynamic equilibrium corresponds to the macrostate that is realized through the maximum number of microstates.
- The thermodynamic entropy of the isolated macroscopic system, coincides with the Gibbs entropy $S(\bar{\mathbf{m}})$, which approximates, in turn, the Shannon entropy of the system.

The statistical mechanics approach puts on a formal basis the idea that thermodynamic equilibrium is simply the macroscopic state of the system that is realized with maximum probability. The fact that the probabilities of the macrostates are obtained, through Eq. (5.10), from a uniform PDF (the microcanonical), stands at the basis of the choice of a uniform partition in the definition of entropy (see discussion in Sect. 2.4). Only in this way, can the statement of maximum probability translate into one of maximum entropy. The uniform nature of ρ^M, descends in turn from incompressibility of the flow in Γ-space, which is consequence of the Hamiltonian character of the microscopic dynamics. In other words, of its conservative character.

In some cases, the subdivision of the system in macroscopic parts, determined by the macrostate $\mathbf{m}$, will correspond to an actual subdivision in subsystems, each containing N_a molecules. In this case, the Γ-space decomposes into a product of subspaces Γ_a, each described by a PDF $\rho_a(\Gamma_a)$. Of course, such an identification is not possible if the parts of the system can exchange molecules, as in the case of the one-molecule phase space volumes associated with the macrostates $m_a = f(\mathbf{x}_a, \mathbf{v}_a)(\Delta x)^3 (\Delta v)^3$. In the case of real subsystems, their macroscopic nature allows to consider the molecules in each of them, independent from those in the others, and to write

---

[3] To have $P(\bar{m}_a) \simeq 1$, we must have $\Delta m_a \gg \sigma_{m_a}$. We recall that in thermodynamic limit conditions $\sigma_{m_a} \sim \bar{m}_a N_a^{-1/2}$.

$$\rho^M(\Gamma) = \prod_a \rho_a(\Gamma_a); \qquad S(\mathbf{m}) = \sum_a S_a(m_a).$$

An important case is the one in which $m_a \equiv E_a$ is the energy of the subsystem. In this case, $\rho_a(\Gamma_a) \equiv \rho_a^M(\Gamma_a)$, and Eq. (5.11) gives us immediately

$$S_a(E_a) = \ln[\Delta\Gamma_a(E_a)/(N_a! \delta\Gamma_a)] \equiv \ln[\Delta\hat{\Gamma}_a(E_a)]. \tag{5.15}$$

(Notice that we recover again the identification of the Boltzmann entropy, with the logarithm of the number of microstates with which the given macrostate is realized).

5.3 The Maximum Entropy Method

It is a technique of rather general nature, for the determination of probability distributions that satisfy a given set of conditions. Imposing that entropy be maximum, guarantees that the only information content in the distribution is the one coming from the constraints.

In statistical mechanics, maximum entropy is what fixes the equilibrium values of the macroscopic parameters. In some cases, e.g. with the choice $m_a = f(\mathbf{x}_a, \mathbf{v}_a)(\Delta x)^3(\Delta v)^3$, such macroscopic parameters can be interpreted as a (macroscopic) probability distribution. The constraints are in this case those provided by conservation of the total energy and number of particles in the isolated system.

We shall use the maximum entropy approach to provide a derivation of the Maxwell-Boltzmann distribution, alternative to that of kinetic theory. Contrary to the kinetic theory approach of Sect. 3.5.1, the one based on the maximum entropy method will allow to bypass any consideration on the microscopic dynamics of the system. This means, in the first place, that no information on the nature of the interactions between the molecules (the form of the collision integral) is required. Perhaps even more important, the fact that the Maxwell-Boltzmann distribution can be generalized to the case of a generic dependences of the molecule energies on the variables $\{p_i, q_i\}$.

In order to proceed, let us subdivide the energy domain for an individual particle in intervals $[\epsilon_k, \epsilon_{k+1}]$, $\epsilon_k = k\Delta\epsilon$. The system macrostate will be determined therefore by the occupation numbers N_k of each interval: $\mathbf{m} \equiv (N_1, N_2, \ldots)$.

To determine the number of microstates corresponding to the macrostate $\mathbf{m}$, we must know the number G_k of cells $(\delta x \delta v)^3$ of one-particle phase space in each energy interval. In the dilute regime $G_k \gg N_k$, in which we are interested, each of these cells will contain at most one particle. Thus, a microstate of the system will be determined once we know which of these cells is occupied (see Sect. 3.4.1). The total number of microstates with which the macrostate $(N_1, N_2, \ldots)$ is realized, will be therefore the product, over all the energy intervals $[\epsilon_k, \epsilon_{k+1}]$, of the ways N_k cells with one molecule, will distribute among G_k available cells. The contribution from each energy interval will be

$$\Delta\hat{\Gamma}(N_k) = \frac{G_k(G_k - 1)\ldots(G_k - N_k)}{N_k!} \simeq \frac{G_k^{N_k}}{N_k!}, \tag{5.16}$$

where use has been made of the condition $G_k \gg N_k$. The corresponding value of the Boltzmann entropy will be

$$S(\{N_k\}) = \sum_k S_k(N_k); \qquad S_k(N_k) = \ln\Delta\hat{\Gamma}_k(N_k). \tag{5.17}$$

Substituting Eq. (5.16) into Eq. (5.17) and using the Stirling approximation $\ln N_k! \simeq N_k(\ln N_k - 1)$, we obtain

$$S(\{N_k\}) \simeq -\sum_k N_k[\ln(N_k/G_k) - 1]. \tag{5.18}$$

To determine the equilibrium distribution, for given values of the total energy E and of the total number of molecules N in the system, we must calculate the maximum in Eq. (5.18) with respect to the N_k's, under the two combined constraints

$$\sum_k N_k = N \quad \text{and} \quad \sum_k N_k\epsilon_k = E. \tag{5.19}$$

The conditional maximum condition is expressed, introducing Lagrange multipliers α and β, in the form

$$dS - \alpha dN - \beta dE = 0.$$

Using Eqs. (5.18) and (5.19) we obtain

$$\frac{\partial}{\partial N_k}(S - \alpha N - \beta E) = -\ln(N_k/G_k) - \alpha - \beta\epsilon_k = 0,$$

which gives us the result

$$N_k = G_k \exp(-\alpha - \beta\epsilon_k). \tag{5.20}$$

The values of the Lagrange multipliers α and β can be determined, at this point, imposing explicitly $\sum_k N_k = N$ e $\sum_k N_k\epsilon_k = E$. It is more interesting to notice the connection between the extremum condition $dS - \alpha dN - \beta dE$, and the expression of the differential in the internal energy E, in terms of the extensive variable S and N: $dE = TdS + \mu dN$. Utilizing these relations in Eq. (5.20), we find immediately

$$N_k = G_k \exp\left(\frac{\mu - \epsilon_k}{T}\right) = \exp\left(\zeta_k + \frac{\mu - \epsilon_k}{T}\right), \tag{5.21}$$

and we notice the relation between the number G_k of cells of Γ-space in the interval $[\epsilon_k, \epsilon_{k+1}]$, and the constant (of the type of the chemical constant) ζ_k.

One question that comes natural at this point, is whether a relation exists between the (statistical mechanics) Boltzmann entropy, Eq. (5.18), and the (kinetic) Boltzmann entropy, Eq. (3.25). Both of them, after all, refer to the distribution of the occupation numbers of energy levels in a system of molecules. We see that the correspondence exists in the thermodynamic limit, $N_k \gg 1$, in which the Boltzmann entropy is described by Eq. (3.23). Writing

$$G_k = \left(\frac{\Delta x \Delta v}{\delta x \delta v}\right)^3 \quad \text{and} \quad N_k = \tilde{f}_{\Delta\epsilon}(\Delta x \Delta v)^3,$$

where $(\Delta x \Delta v)^3$ is the volume of one-molecule phase space corresponding to the energy interval $[\epsilon_k, \epsilon_{k+1}]$, we find immediately, combining with Eq. (5.18):

$$S = -\sum_k N_k \ln(N_k/(eG_k)) \simeq - \int d^3x d^3v \; \tilde{f}_{\Delta\epsilon} \ln(\tilde{f}_{\Delta\epsilon}(\delta x \delta v)^3/e), \qquad (5.22)$$

that is equal, within constants, to the expression of the Boltzmann entropy provided by Eq. (3.25).

5.3.1 From Classical to Quantum Statistics

Quantum mechanics fixes a microscale $\delta\Gamma$, below which, different points in the Γ-space of a system are physically indistinguishable. In the case of a gas of N molecules: $\delta\Gamma \sim \hbar^{3N}$. This induces naturally a subdivision of the Γ-space in cells corresponding to individual quantum states of the system. The choice of the quantum states (the shape of the cells) is in general arbitrary. For the kind of problems in which we are interested, however (stationary statistics of an isolated system), the natural choice is that of the energy eigenstates, that are in fact the stationary states for the system. To fix the ideas, consider the case of a gas of N non-interacting particles, placed in a harmonic one-dimensional potential well. Let us indicate with $|a; k\rangle$ the kth eigenstate of the ath particle; the corresponding energy eigenvalue will then be $E_k = (1/2 + k)\hbar\omega$, where ω is the classical oscillation frequency of the particle in the potential.

How to pass from the one-particle, to the N-particle states, as usual, turns out to be the difficult part of the problem.

In a classical perspective, the Γ-space of the gas, would be the cartesian product of the phase-spaces of the particles; therefore, $\Gamma = (n_1, n_2, \ldots n_N)$ would be the vector of the states n_a of the particles $a = 1, 2, \ldots N$. The fact that states of the system that differ only by a particle permutation, e.g. $\Gamma = (n_1, n_2, \ldots, n_N)$ and $\Gamma' = (n_2, n_1, \ldots, n_N)$, are physically indistinguishable, advised us to consider as microstates of the system, not the individual cells, rather, their unions under permu-

Fig. 5.1 Possible states of two identical particles a and b. Classically, α and β are indistinguishable states. Together they form a microstate of weight 2, while state γ has weight 1

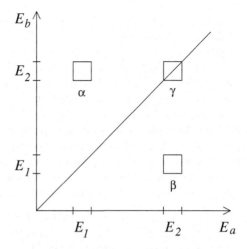

tation of the components. Always working half way between classical and quantum, the statistical weight of microstates in which all the particles have different energies, would be larger than that of those in which some of the energies are equal. Classically, the statistical weight of the microstate is its volume, that will be proportional to the multinomial factor $N!/ \prod_k N_k!$, where N_k is the number of particles with energy E_k. The situation is illustrated in Fig. 5.1, in the case $N = 2$: the two cells α and β form together the microstate "one particle with energy E_1 and one with energy E_2". This microstate has clearly a statistical weight that is twice that of a state in which the particles have identical energy.

In a fully quantum description, the state of an N-particle system is described by combinations of wave functions of the particles in the system, that will be symmetric or antisymmetric, depending on whether the particles are bosons (Bose-Einstein statistics) or fermions (Fermi-Dirac statistics). In the case $N = 2$:

$$|E_1, E_2\rangle = \frac{1}{\sqrt{2}}[|a; 1\rangle|b; 2\rangle \pm |a; 2\rangle|b; 1\rangle]. \tag{5.23}$$

In other words, α and β in Fig. 5.1, correspond to the same quantum state, and not to two indistinguishable states, which, put together, weigh 2. In case of Bose-Einstein statistics, state $|E_1, E_2\rangle$, with $E_1 \neq E_2$, weighs as much as $|E_2, E_2\rangle = |a, 2\rangle|b, 2\rangle$; in the case of fermions, the state $|E_2, E_2\rangle$ simply does not exist (that is the content of the **Pauli exclusion principle**). The quantum state of an N-particle system can therefore be written, both in the case of fermions and of bosons, in the form $|\Gamma\rangle \equiv |N_1, N_2, \ldots\rangle$, where N_a indicates the number of particles with energy E_a, and all these states will have identical statistical weight (the normalization of the state) equal to one. The final step, from quantum mechanics to quantum statistical mechanics, is accomplished disregarding the phase of the wave function $|\Gamma\rangle$, working therefore with simple probabilities (density matrix description), and not with amplitudes of the states.

The difference between classical and quantum statistics is made clear, in the case of fermions, by the Pauli principle. In the case of bosons, the difference is more subtle, and has to do with the way the many-particle states are weighed. The fact that cells corresponding to particles with different energy, weigh the same as "degenerate" states with identical energy particles, means that degenerate states are more likely than what would be expected, if Boltzmann statistics continued to be valid for non-dilute systems. Now, Eq. (5.21) tells us that in the classical case, the low energy states are the ones that contain more particles. Thus, low energy states will tend to be more occupied in the case of bosons than in the hypothetical case of non-dilute Boltzmann particles. The opposite in the case of fermions. At low temperatures, the energy content of a Bose gas will be lower than that of an identical system of Boltzmann particles. The opposite, again, in the case of fermions.

Note We have obtained the equilibrium distribution $\{\bar{N}_k\}$, in the case of a dilute gas, maximizing the entropy $S = \ln(\Delta\hat{\Gamma}(\{N_k\}))$, where $\Delta\hat{\Gamma}(\{N_k\})$ was the number of microstates with which the macrostate $\{\bar{N}_k\}$ was realized. From a purely classical perspective, the physical meaning of this maximization process was to find the most probable macrostate, i.e. the one occupying the biggest portion of $\Delta\Gamma(E)$. It is therefore clear that the choice of partition of $\Delta\Gamma(E)$ must be irrelevant in the determination of the most probable macrostate. In particular, a partition that considers particles distinguishable (the one that would lead to the definition of entropy in Eq. (3.17)), should produce the same result as one that considers particles indistinguishable.

If particles are considered distinguishable, the number of microstates will be simply the number of ways the N_k particles in the interval $[\epsilon_k, \epsilon_{k+1}]$, can be distributed among the G_k cells $(\delta x \delta v)^3$ available, for all k. There are $N!/\prod_k N_k!$ ways to group N particles in sets of N_k. Each of the N_k particles in the interval $[\epsilon_k, \epsilon_{k+1}]$ has then G_k cells out of which to choose. The end result is the multinomial product

$$\Delta\hat{\Gamma}^{dist}(\{\bar{N}_k\}) = N! \prod_k \frac{G_k^{N_k}}{N_k!} = N!\Delta\hat{\Gamma}(\{\bar{N}_k\}), \qquad (5.24)$$

where $\Delta\hat{\Gamma}(\{\bar{N}_k\}) = \prod_k \Delta\hat{\Gamma}(\bar{N}_k)$ is the number of states in the indistinguishable case, described by Eq. (5.16). We can then maximize the Boltzmann entropy $S^{dist}(\{\bar{N}_k\}) = \ln \Delta\hat{\Gamma}^{dist}(\{\bar{N}_k\})$, following the same exact procedure in the previous section. The result will be again the Boltzmann distribution Eq. (5.21), independently of the fact that the gas is dilute or not. We have seen however that this cannot be true: we are counting states of the system that not only are classically indistinguishable, but that also, from the point of view of quantum mechanics, do not exist.∎

5.4 The Gibbs Distribution

We have seen in Sect. 5.2, that a central role in the derivation of the thermodynamic properties of a macroscopic system, is played by the way in which the system is subdivided. After having considered in Sect. 5.3 the case of a subdivision in energy shells, we consider now that of a subdivision in physically distinct macroscopic parts, something that we have already seen at work when studying the properties of a thermodynamic system in a heat bath. In fact, what we want to determine now is precisely the distribution in the Γ-space of a system in equilibrium with a thermal bath. Contrary to the case of the microcanonical, that was referred to an isolated system, the new distribution, called **canonical** or **Gibbs distribution**, will be able to describe the energy fluctuations in the system.

Proceeding as in Sect. 4.7, let us consider an isolated macrosystem, composed by a system and a much larger heat bath, that can exchange heat, but whose volumes and number of particles remain otherwise invariant. In the same notation of Sect. 4.7, we shall identify variables referred to the heat bath and to the composite macrosystem, with subscripts zero and tot, respectively, while variables referred to the system will carry no subscript.

Calculations are simpler working with number of microstates, rather than with volume elements of Γ-space. We thus indicate with $\hat{\Gamma}$ the index of the microstate Γ, in such a way that

$$d\hat{\Gamma} = (N!\delta\Gamma)^{-1}d\Gamma.$$

Given the subdivision of the macrosystem in parts, each with a well defined set of particles, the Γ space of the macrosystem will be the product of the Γ spaces of the two parts. We can write therefore, for the marginal distribution in the Γ-space of the system:

$$\rho(\hat{\Gamma}) = \int \rho^M(\hat{\Gamma}_{tot})d\hat{\Gamma}_0 = \frac{\Delta\hat{\Gamma}_0(\Gamma)}{\Delta\hat{\Gamma}_{tot}(E_{tot})}, \tag{5.25}$$

where $\Delta\hat{\Gamma}_0(\Gamma)$ is the number of microstates in the heat bath corresponding to the microstate Γ in the system.

Now, if both the system and the heat bath are macroscopic, a macroscopic property of the second, such as the number of microstates $\Delta\hat{\Gamma}_0(\Gamma)$ (associated with the Boltzmann entropy of the heat bath), will not depend on the microstate Γ, except for the macrostate of the system it determines. In our case, $E(\Gamma)$. In other words, $\Delta\hat{\Gamma}_0(\Gamma)$ is determined by the macrostate of the heat bath, that results from the fact that the energy of the system is E. This is just the energy of the heat bath, $E_0 = E_{tot} - E(\Gamma)$. We have therefore that the marginal PDF $\rho(\Gamma)$ is uniform for each value of E, and we can define

$$\rho(\hat{\Gamma}) = \frac{\Delta\hat{\Gamma}_0(E_{tot} - E(\Gamma))}{\Delta\hat{\Gamma}_{tot}(E_{tot})}. \tag{5.26}$$

Exploiting Eq. (5.15), the volume $\Delta\Gamma_0(E_0)$ can be written as a function of the Boltzmann entropy. We find finally

$$\rho(\hat{\Gamma}) = \frac{1}{\Delta\hat{\Gamma}_{tot}(E_{tot})} \exp(S_0(E_{tot} - E(\Gamma))) \simeq \frac{1}{Z} \exp(-\frac{E(\Gamma)}{T}), \tag{5.27}$$

where use has been made of the relation $S_0(E_{tot} - E) \simeq S_0(E_{tot}) - E S_0'(E_{tot}) = S_0(E_{tot}) - E/T$, T is the temperature of the thermal bath, and $Z = \Delta\hat{\Gamma}_{tot}(E_{tot})/\Delta\hat{\Gamma}_0(E_{tot})$ is a normalization factor, called the **partition function**.

The PDF ρ in Eq. (5.27), is called **Gibbs distribution** (or **canonical distribution**). Although $\rho(\hat{\Gamma})$ can be expressed through Eq. (5.27) as a function solely of E, it is clear that it is not the PDF for E. In fact, it is not peaked at all at the equilibrium value $\bar{E} \simeq NT$, instead, the maximum of $\rho(\Gamma)$ is at $E = 0$. The peak at $\bar{E} \simeq NT$ in the macroscopic PDF $\rho(E)$, is produced by the volume factor in the change of variables from $\Gamma \in \Delta\Gamma(E)$ to E:

$$\rho(E) = \rho(\hat{\Gamma})\frac{\Delta\hat{\Gamma}(E)}{\Delta E} = \frac{1}{Z\Delta E} \exp\left(\frac{TS(E) - E}{T}\right) \qquad (5.28)$$

(for the calculation of $\Delta\Gamma(E)$, we have used again Eq. (5.15)). We recognize in $-TS(E) + E = \hat{F}_T(S(E))$ the free energy content of the non-equilibrium state, introduced in Sect. 4.7. We see at once that $\rho(E)$ has a peak of width $\Delta E \sim T$ in correspondence to the minimum of $\hat{F}_T(S(E))$. This minimum is determined by the condition

$$\frac{1}{T} = \frac{\partial S}{\partial E},$$

that is satisfied only if E is equal to its equilibrium value in a thermal bath at temperature T. We find therefore that the minimum of $TS(E) - E$, with respect to E, is nothing but the Helmholtz free energy $F(T, V) = \bar{E} - TS(\bar{E}, V) \equiv E(T, V) - TS(T, V)$.

5.4.1 Role of the Partition Function

The connection between Helmholtz free energy and canonical distribution has some important consequences. Such consequences become clear when it comes to calculating the partition function Z. Imposing the normalization condition $\int \rho_A(E_A)\,\Gamma_A = 1$, we find immediately, using Eq. (5.27):

$$Z = \int d\hat{\Gamma} \exp\left(-\frac{E_A(\Gamma)}{T}\right), \qquad (5.29)$$

and we see that the integral is dominated by the equilibrium configuration for $\bar{E}$. (We recall that the integral in Eq. (5.29) is actually a sum over microstates and the continuum limit is valid only in the case of a classical system). We can write therefore

$$Z \simeq \Delta\hat{\Gamma}(\bar{E}) \exp\left(-\frac{\bar{E}}{T}\right) = \exp\left(\frac{TS(\bar{E}) - \bar{E}}{T}\right),$$

where use has been made, once more, of Eq. (5.15). We have already seen that $\bar{E} - TS(\bar{E}) \equiv E(T, V) - TS(T, V)$ is the Helmholtz free energy for the subsystem in a thermal bath at temperature T. We find therefore the fundamental relation

$$F(T, V) = -T \ln Z = -T \ln \left[\int d\hat{\Gamma} \exp \left(-\frac{E(\Gamma)}{T} \right) \right], \qquad (5.30)$$

that provides us with a direct method to calculate the free energy of a system from an average in Γ space.

It could be worth at this point to repeat the calculation of the Helmholtz free energy of an ideal gas, using the partition function method, from Eq. (5.30). We have in this case $E = (2m)^{-1} \sum_i p_i^2$, so that

$$N! \delta \Gamma Z = \int d^{3N}q d^{3N}p \, \exp \left(-\frac{1}{2mT} \sum_{i=1}^{3N} p_i^2 \right) = V^N \left[\int dp \exp \left(-\frac{p^2}{2mT} \right) \right]^{3N}$$

$$= V^N (2\pi mT)^{\frac{3N}{2}} \exp(N(\ln V + \frac{3}{2} \ln T + \frac{3}{2} \ln(2m\pi))). \qquad (5.31)$$

Combining with Eq. (5.30), and using the Stirling formula $\ln N! \simeq N(\ln N - 1)$, we find in the end

$$F(T, V, N) = NT \left(\frac{3}{2}(1 - \ln T) - \ln(V/N) - \zeta \right),$$

where we have exploited the relation $\delta \Gamma = (\delta p \delta q)^{3N}$, to write $\zeta = 5/2 - (3/2) \ln(\delta q^2 \delta p^2/(2\pi m))$ (compare with Eq. (4.67)).

5.5 The Grand Canonical Distribution

The idea of the Gibbs distribution could be generalized considering a subdivision of the system in subsystems that, beyond energy, can also exchange particles. The distribution function of the system, called the **grand canonical distribution**, will now be able to account both for the fluctuations in the energy and in the number of particles in the system. The thermal bath will now be also a "particle bath" at chemical potential μ, that will determine the equilibrium value of the particle number in the system. We expect therefore that some variation of the partition function method should allow determination of the thermodynamic properties of the system, including the equilibrium values of the particle numbers. As we shall see, a central role in the procedure, analogous to that of $F(T, V)$ in the canonical case, will be played here by the thermodynamic potential $\Omega(T, V, \mu)$, introduced in Sect. 4.8.3.

The procedure to derive the grand canonical distribution, follows closely that for the canonical, the main difference being that, due to particle exchange, we cannot write the Γ-space of the isolated macrosystem, as the product of the Γ-spaces of its parts. The system microstate, Γ, is now a vector with a variable number of components, that depends on the number of particles in the system. Nevertheless, the form of Eq. (5.25) will remain the same, the only modification being the domain of the integral $\int d\hat{\Gamma}_0$, that will now be a function of N. Similarly, Eq. (5.26) maintains its form, but the number of states $\Delta\hat{\Gamma}_0$ will now be a function also of N_0:

$$\rho(\hat{\Gamma}) = \frac{\Delta\hat{\Gamma}_0(E_{tot} - E, N_{tot} - N)}{\Delta\hat{\Gamma}_{tot}(E_{tot})}.$$

In analogy with Eq. (5.27), we can therefore write

$$\rho(\hat{\Gamma}) = \frac{1}{\Delta\hat{\Gamma}_{tot}(E_{tot})} \exp(S_0(E_{tot} - E, N_{tot} - N)).$$

We expand $S_0(E_{tot} - E, N_{tot} - N)$ around (E_{tot}, N_{tot}), exploiting Eq. (4.55), that can be rewritten in the form, considering constant volume $(dV = 0)$:

$$dS = \frac{dE}{T} - \frac{\mu}{T}dN.$$

The grand canonical distribution will read

$$\rho(\hat{\Gamma}) = \frac{1}{\mathscr{Z}} \exp\left(\frac{\mu N - E}{T}\right), \tag{5.32}$$

where the normalization function $\mathscr{Z} = \Delta\hat{\Gamma}_{tot}(E_{tot})/\Delta\hat{\Gamma}_0(E_{tot}, N_{tot})$ is called the **grand partition function**.

As in the case of the partition function in the canonical distribution, the grand canonical partition function can be calculated from thermodynamic considerations. In analogy with Eq. (5.29), we can write

$$\mathscr{Z} = \sum_N \int d\hat{\Gamma}_N \exp\left(\frac{\mu N - E}{T}\right), \tag{5.33}$$

where the notation $d\hat{\Gamma}_N$ indicates that the integral is performed for a fixed number of particles. Considering that the integral in Eq. (5.33), is dominated by equilibrium configurations, we find

$$\mathscr{Z} \simeq \Delta\hat{\Gamma}(\bar{E}, \bar{N}) \exp\left(\frac{\mu\bar{N} - \bar{E}}{T}\right) = \exp\left(\frac{TS(\bar{E}, \bar{N}) + \mu\bar{N} - \bar{E}}{T}\right).$$

The expression $-TS(\bar{E}, \bar{N}) - \mu\bar{N} + \bar{E}$, however, is just the thermodynamic potential $\Omega(T, V, \mu)$, defined in Eq. (4.64). Thus

$$\mathscr{Z} = \exp(-\Omega(T, V, \mu)/T),$$

and the grand canonical distribution becomes

$$\rho(\hat{\Gamma}) = \exp\left(\frac{\Omega + \mu N - E}{T}\right).$$

The grand canonical distribution allows to express the thermodynamic potential Ω in function of the microscopic properties of the system. In analogy with Eq. (5.30), we can write in fact

$$\Omega(T, V, \mu) = -T \ln \mathscr{Z} = -T \ln\left[\sum_N \int d\hat{\Gamma}_N \, \exp\left(\frac{\mu N - E}{T}\right)\right]. \qquad (5.34)$$

The microcanonical, the canonical and the grand canonical are the three fundamental distributions of statistical mechanics. They focus on different characteristics of an isolated thermodynamic system, but, for the rest, they are equivalent. In most situations, the fluctuations in the energy and in the number of particles in macroscopic parts of the system are negligible, and the three distributions predict identical behaviors for the thermodynamic variables. The main advantage of the canonical and the grand canonical distribution lies therefore more in the ease of calculation, and in the power of instruments such as the partition and grand partition function, than else.

5.5.1 Application: Bose-Einstein and Fermi-Dirac Statistics

The grand canonical version of the Gibbs distribution approach, does not require a partition of the system in subvolumes, as in the canonical case. We can in particular consider the subdivision in energy shells, that has been utilized to derive the Boltzmann distribution in Sect. 5.3. We illustrate here an application of this strategy to a gas of non-interacting quantum particles, in the two cases of Bose-Einstein and Fermi-Dirac statistics.

Proceeding as in Sect. 5.3, we partition the system in one-particle energy intervals $[\epsilon_k, \epsilon_{k+1}]$, indicating with N_k the number of particles in each interval. In analogy with Eq. (5.34), the thermodynamic potential Ω corresponding to one of these intervals, will be in the form

$$\Omega_k = -T \ln \mathscr{Z}_k = -T \ln\left[\sum_{N_k} \int d\hat{\Gamma}_{k,N_k} \, \exp\left(\frac{\mu N_k - E_k}{T}\right)\right],$$

where, from the fact that the particles are non-interacting, we can write $E_k = N_k \epsilon_k$. Thus

$$\Omega_k = -T \ln\left[\sum_{N_k} \int d\hat{\Gamma}_{k,N_k} \, \exp\left(\frac{(\mu - \epsilon_k)N_k}{T}\right)\right], \qquad (5.35)$$

The equilibrium value of the occupation numbers of the energy interval is then obtained from

$$\bar{N}_k = \frac{\partial \Omega_k}{\partial \mu}. \tag{5.36}$$

We see that the term $\exp\{(\mu - \epsilon_k)N_k/T\}d\hat{\Gamma}_{k,N_k}$ in Eq. (5.35) factors into contributions from individual one-particle quantum states:

$$d\hat{\Gamma}_{k,N_k} = \prod_j dn_{j,k}, \qquad \sum_j n_{j,k} = N_k,$$

with $n_{j,k}$ the number of particles occupying the jth state in the energy interval $[\epsilon_k, \epsilon_{k+1}]$. The thermodynamic potential Ω_k takes then the form

$$\Omega_k = -T G_k \ln \left[\sum_n \exp \left(\frac{(\mu - \epsilon_k)n}{T} \right) \right], \tag{5.37}$$

where G_k is the number of one-particle states in the interval $[\epsilon_k, \epsilon_{k+1}]$.

We must now distinguish between Fermi-Dirac and Bose-Einstein statistics. In the first case, the Pauli exclusion principle restricts the range of occupation numbers of the quantum states to $n = 0, 1$; in the case of bosons, all occupation numbers are allowed.

In the Fermi-Dirac case, we find

$$\Omega_k = -T G_k \ln \left[1 + \exp \left(\frac{\mu - \epsilon_k}{T} \right) \right], \tag{5.38}$$

which, substituting into Eq. (5.36), gives us the equilibrium distribution

$$\bar{N}_k = \frac{G_k}{1 + \exp\{(\epsilon_k - \mu)/T\}}. \tag{5.39}$$

This is the equilibrium distribution for a gas of fermions: the so-called **Fermi-Dirac distribution**. As in the Boltzmann case, the chemical potential μ is determined from the normalization condition $\sum_k N_k = N$. We see that at large energies, $\exp\{(\epsilon_k - \mu)/T\} \gg 1$, and the Boltzmann distribution is recovered.

In the Bose-Einstein case, the geometric sum in the logarithm in Eq. (5.37) must be carried out up to $n_k = \infty$, and the result is

$$\Omega_k = T G_k \ln \left[1 - \exp \left(\frac{\mu - \epsilon_k}{T} \right) \right]. \tag{5.40}$$

Substituting into Eq. (5.36), gives us this time

$$\bar{N}_k = \frac{G_k}{1 - \exp\{(\epsilon_k - \mu)/T\}}, \tag{5.41}$$

that is the equilibrium distribution for a gas of free bosons: the so-called **Bose-Einstein distribution**. As in the Fermi-Dirac case, the Boltzmann distribution is recovered at high energies, for which $\exp\{(\epsilon_k - \mu)/T\} \gg 1$. We notice the property, already discussed in Sect. 5.3.1, that low energy states are more likely to be occupied in the case of bosons, than in that of fermions, with the case of particles obeying Boltzmann statistics, lying in the middle.

5.6 The Equipartition Theorem

The techniques of statistical mechanics can be applied to any system whose Hamilton function $\mathcal{H}$ is known. The only condition that must be satisfied, in order for the Gibbs formalism to be applicable, is that $\mathcal{H}(\Gamma) > -\infty$ in all the microstates Γ.[4]

The case in which the Hamiltonian is quadratic in at least part of the variables, is specially important, because the contribution from these variables to the internal energy of the system, can be calculated analytically in terms of Gaussian integrals. A typical situation is that of an Hamiltonian that is quadratic in the momenta:

$$\mathcal{H} = \sum_{i=1}^{N} \frac{p_i^2}{2m_i} + U(\{q_i, i = 1, \ldots N\}). \tag{5.42}$$

Notice that we do not assume the mass coefficients to be all equal, as we may have different kinds of molecules in the system, as well as rotational contributions to the kinetic energy (in which case the mass coefficient would be a moment of inertia).

The internal energy of a system at temperature T can be calculated from the canonical distribution Eq. (5.27). For a classical system for which $\delta\Gamma \to 0$, we can write

$$\bar{E} = Z^{-1} \int d\hat{\Gamma} \, \mathcal{H}(\Gamma) \exp(-\mathcal{H}(\Gamma)/T). \tag{5.43}$$

In the case of a Hamiltonian in the form of Eq. (5.42), the integral in Eq. (5.43) factorizes in contributions from the different moments, and the one from the coordinates. Similarly for the partition function, that can be written in the form

$$Z = (N! \delta\Gamma)^{-1} Z_{\mathbf{q}} \prod_{k} Z_{p_k},$$

where

$$Z_{\mathbf{q}} = \int \prod_{k} dq_k \, \exp(-U/T) \quad \text{and} \quad Z_{p_k} = \int dp_k \exp(-p_k^2/(2m_k T)).$$

---

[4] In quantum mechanics, this means that the spectrum of $\mathcal{H}$ is bounded from below.

The Z_{p_k} are the normalizations for the integrals over the p_k's; $Z_{\mathbf{q}}$ is the normalization for the multiple integral over the q_i's. Substituting into Eq. (5.43), we see that each momentum contributes to E a term

$$\bar{E}_{p_k} = Z_{p_k}^{-1} \int dp_k \frac{p_k^2}{2m_k} \exp(-\frac{p_k^2}{2m_k T}) = \frac{T}{2}, \tag{5.44}$$

independent of the mass coefficient m_i. Thus, the internal energy of the system is

$$\bar{E} = \frac{NT}{2} + \langle U \rangle, \tag{5.45}$$

which tells us that the kinetic interpretation of temperature will continue to be valid also in the presence of interactions. The only condition that has to be verified, is that the kinetic part of the Hamiltonian has the standard quadratic form. In this case, each momentum will contribute $T/2$ to the internal energy, and it is clear that an identical contribution will be produced by any other coordinate q_i on which $\mathscr{H}$ depended quadratically. This is the content of the **equipartition theorem**:

• Any quadratic degree of freedom in the dynamics of a thermodynamic system contributes $T/2$ to its internal energy.

In the case of an ideal gas, in which $U = 0$, we see that each molecule will contribute $T/2$ for each linear momentum component. This produces the standard expression for the internal energy of a monoatomic gas $E = \frac{3}{2}NT$.

5.7 The Ising Model

Up to now, we have considered only thermodynamic systems whose microscopic dynamics is continuous. There is a whole set of systems, in which the microscopic dynamics is—or can be treated—as discrete.

The Ising model is an example of systems, in which the microscopic dynamics is intrinsically discrete. It is a model of ferromagnetism, in which the discrete variables represent the magnetic dipoles of the atomic spins, arranged on a lattice. The spins can take only two values, $s = \pm 1/2$, but extensions of the model exist, in which the spin variables are allowed to take continuous values (see Table 5.1). Each spin is allowed to interact only with its neighbors, but again there are variations of the model, in which the interactions decay more slowly. The two-dimensional Ising model is one of the simplest models of statistical mechanics in which a phase transition is present.

The Ising model statistics is described by the Hamiltonian

$$\mathscr{H}[\sigma] = -J \sum_{<i,j>} \sigma_i \sigma_j - \mu H \sum_i \sigma_i, \tag{5.46}$$

Table 5.1 Notable spin models

N-vector model	$\mathcal{H} = -J \sum_{<i,j>} \sigma_i \cdot \sigma_j$, with $\sigma_i \in \mathbf{R}^n$, $|\sigma| = 1$; dimension of the lattice generic
Heisenberg model	Version of the n-vector model with $n = 2$
XY model	Version of the n-vector model with $n = 3$
Potts model	Version of the XY model, in which $\sigma = (\cos\theta, \sin\theta)$ takes discrete values: $\theta = 2\pi n/m$, $n = 1, 2, \ldots, m$
Edwards-Anderson model	Ising model on a "disordered" lattice: $\mathcal{H} = -\sum_{<i,j>} J_{ij}\sigma_i\sigma_j$, with J_{ij} random
Sherrington-Kirkpatrik model	Version of the Edwards-Anderson model with long-range random interactions

where $\sigma_i = \pm 1$, is the spin variable in site i, the notation $<i, j>$ indicates that the sum is carried out only on nearest neighbors, and we have admitted the possibility of interaction with an external uniform magnetic field H, with μ indicating the electron magnetic moment. We notice that $\mathcal{H}[\sigma]$ is a purely static Hamiltonian, with no variables which can be interpreted as a velocity.

Knowing the Hamiltonian $\mathcal{H}[\sigma]$, the canonical formalism can be applied to obtain the distribution of the states of the system in a thermal bath at temperature T:

$$\rho[\sigma] = \frac{1}{Z} \exp(-\frac{\mathcal{H}[\sigma]}{T}), \qquad Z = \sum_\sigma \rho[\sigma], \qquad (5.47)$$

where $\sum_\sigma \equiv \sum_{\sigma_1} \sum_{\sigma_2} \ldots$ indicates sum over all spin configurations on the lattice. Proceeding as in the case of Eq. (5.28), we can obtain the distribution of the energies $\mathcal{H}[\sigma] = E$:

$$\rho(E) = \frac{1}{Z\Delta E} \exp\left(\frac{TS(E) - E}{T}\right) \qquad (5.48)$$

Thermal equilibrium corresponds to the minimum with respect to E of the free energy $\hat{F}(E) = E - TS(E)$: $F(T) = \hat{F}(E(T))$.

We shall consider a cubic lattice, so that $i = (i_x, i_y, i_z)$ determines the site coordinate $\mathbf{x}_i = (i_x\Delta, i_y\Delta, i_z\Delta)$, with Δ the lattice constant. For a cubic lattice, the nearest neighbors of site i will be $j = (i_x \pm 1, i_y, i_z)$, $(i_x, i_y \pm 1, i_z)$, $(i_x, i_y, i_z \pm 1)$.

In the absence of an external magnetic field, we see that, for $J > 0$, a configuration with all spins having identical sign is energetically favored, while, for $J < 0$, the opposite occurs. The first condition describes the behavior of a ferromagnetic material. The second, that of an antiferromagnetic material: the lowest energy state is that in which adjacent spins have opposite sign.

Ferromagnetic materials are known to undergo a second order phase transition at a critical temperature T_C, called the **Curie temperature**, above which the material becomes paramagnetic.

As already mentioned, the Ising model is able to predict the presence of a phase transition in a ferromagnetic material. This phase transition is present only in dimension two or higher, the one-dimensional Ising model being paramagnetic for all $T > 0$. The paramagnetic properties of the one-dimensional Ising model, and the presence of a ferromagnetic phase transition in the two-dimensional Ising model, can be explained utilizing a simple argument, due R. Peierls:

- In the ferromagnetic phase, the creation of an island of equal sign adjacent spins, in a uniform sea of spins with opposite sign, should be disfavored. The island should rapidly evaporate, leaving the system in a fully homogeneous state. Creation of a island should therefore be associated with an increase of the free energy of the system, with respect to the spatially homogeneous case.
- In the paramagnetic phase, the opposite should occur: the unstable nature of the spatially homogeneous state, should be revealed by the fact that creating an island, will produce a decrease in the free energy of the system.

5.7.1 The 1D Case

Consider a linear chain composed of N spins. Let us impose for simplicity periodic boundary conditions (the chain is closed in a ring). To determine the free energy content, $\hat{F}_T(L) = E(L) - TS(L)$, of an island containing L spins, we must determine the energy $E(L)$ required to create such island. We also need its Boltzmann entropy $S(L)$, that is the logarithm of the number of states corresponding to presence of the island in the system.

The situation is illustrated in Fig. 5.2a: creating the island requires an amount of energy $2J$ for each of the two "kinks" that constitute the island border, where the energy has changed from $-J\sigma_i\sigma_{i+1} = -J$, in the homogeneous state, to $-J\sigma_i\sigma_{i+1} = J$, in the final state (similarly in site j). The energy cost to create the island is therefore

$$\Delta E(L) = 4J, \tag{5.49}$$

independent of the island length.

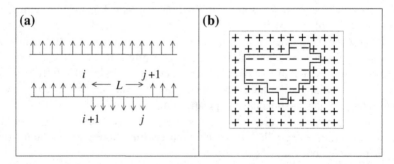

Fig. 5.2 Typical configurations of spin islands in a 1D (*left*) and 2D (*right*) Ising model. In both cases the energy required to create the island is concentrated in the island boundary

To calculate the Boltzmann entropy, we must determine the number of states in the system with and without the island. In the homogeneous state (without the island), we have just two states (all spins up or all down). In presence of the island, we must count all the ways we can create the island from the two homogeneous initial conditions. This is just the number of ways we can choose the location of site i (the other, $j = i + L$ mod. N, is fixed), out of each initial homogeneous state. There are N ways to choose i in each homogeneous state, for a total of $2N$. The entropy gain with respect to the homogeneous state is therefore

$$\Delta S = \ln N. \tag{5.50}$$

Putting together the two contributions, we find for the change in free energy:

$$\Delta \hat{F}_T = \Delta E - T \Delta S = 4J - T \ln N, \tag{5.51}$$

that will be negative in the thermodynamic limit for all $T > 0$. Islands will continue to form, pushing the system away from the spatially homogeneous state. The system is always paramagnetic and no phase transition is present.

5.7.2 The 2D Case

The system is now a square lattice of side N. As in the 1D case, the spins variable can only take values $\sigma_i = \pm 1$, indicated in Fig. 5.2b with $\pm$. To each pair of neighboring sites $<i, j>$, we can associate a link in the lattice and a contribution $-J\sigma_i\sigma_j$ to the energy. In the case of a square lattice, the links between the sites will be arranged themself on a square lattice identical to the one formed by the sites.

As in the 1D case, we see that the energy required to create the island is concentrated in the links that connect sites in which one of the spins has flipped sign. This is again the boundary of the island (see Fig. 5.2b). The energy cost will be therefore

$$\Delta E(L) = 2JL \tag{5.52}$$

where L is the number of links at the boundary (the length of the boundary).

We pass to determine the entropy gain $\Delta S(L)$ associated with creation of the island. The problem is much more complicated than in the case of Eq (5.50). Determining the number of islands of perimeter L, $\Delta \hat{\Gamma}(L)$, out of which the entropy gain in the island formation can be calculated, $\Delta S(L) = \ln(\Delta \hat{\Gamma}(L))$, requires indeed counting all the possible shapes and locations of the islands. The problem is equivalent to counting all the closed non-intersecting random trajectories, connecting L links, that can be drawn in the lattice. Calculating the exact number of such trajectories is going to be very difficult; some estimate can nevertheless be obtained.

We start by observing that this number is going to be, in any case, smaller than the number of random paths that can be drawn, arbitrarily joining adjacent links. For a square lattice, at each step in the random path, the walker will have at its disposal four links to which to connect, one of which was the one in which it was located at the previous step. If we do not count the possibility of tracing back the trajectory, we are left with at most three valid choices at each step. In a sequence of L steps, we will have therefore at our disposal, at most 3^{L-1} ways to select a path from a given initial link. Counting the initial links available in the lattice, we find for the number of ways in which a paths of length L can be drawn in a square lattice of side N:

$$\Delta\hat{\Gamma}(L) < N^2 3^{L-1}. \tag{5.53}$$

A refinement of this estimate can be obtained requiring that the paths do not intersect themselves. This means that, if the walker reaches a link that is adjacent to another link already hit by the trajectory, at the next step, the number of valid choices will be reduced by one. Thus, two possible links where to jump. We can expect that the walker will have available at each step at least two choices, that gives a minimum number of paths on the lattice $N^2 2^{L-1}$. Putting together with Eq. (5.53):

$$N^2 2^{L-1} < \Delta\hat{\Gamma}(L) < N^2 3^{L-1}, \tag{5.54}$$

that gives us for the entropy:

$$L \ln 2 < S(L) - 2 \ln N < 3 \ln L. \tag{5.55}$$

If we consider both L and M macroscopic, we can disregard the logarithmic factors $\ln N$ in front of L. Putting together with Eq. (5.52), we are thus left with

$$\Delta\hat{F}_T = L\left[2J - T \ln C\right], \qquad 2 < C < 3. \tag{5.56}$$

Equation (5.56) tells us that there must be a phase transition at a critical temperature

$$T_c = \frac{2J}{\ln}C, \quad \text{with} \quad 1.820 < \frac{2}{\ln C} < 2.885. \tag{5.57}$$

For $T > T_c$, we have in fact $\Delta\hat{F}_T < 0$, as in the 1D case, corresponding to a paramagnetic regime. For $T < T_c$, $\Delta\hat{F}_T$ changes sign and the system becomes ferromagnetic.

5.8 Problems

Problems 1 Consider a system of four *distinguishable* particles, each of which can lie in energy levels $\epsilon = n\Delta, n = 0, 2, 3, \ldots$

- Determine the microstates that corresponds to the macrostate $E = 6\Delta$. Evaluate the entropy of this state.
- Assuming the energy levels a-priori equiprobable, determine the one-particle distribution $f_1(\epsilon)$ corresponding to the macrostate $E = 6\Delta$.

- Determine the microstates corresponding to $E = 5\Delta$, and use this information, together with that relative to $E = 6\Delta$, to obtain an estimate of the temperature corresponding to $E = 5 \div 6\,\Delta$.

Problems 2 Suppose that a loaded dice had $P(\{1, 2\}) = 1/3$, $P(\{3, 4\}) = 1/2$, $P(\{5, 6\}) = 1/6$, and that the prize for the different result be $x(1) = x(2) = 1$; $x(3) = x(4) = 0$; $x(5) = x(6) = -1$. Hence, $\langle x \rangle = 1/6$ (in case of an unloaded dice, of course, $\langle x \rangle = 0$).

- Calculate, for $N \gg 1$, the probability $P(x|\langle x \rangle_N = 0)$, i.e. the typical distribution of outcomes for x, that would produce a sample average identical to that of the unloaded case. (Seek the analogy with the maximum entropy method in the case of the velocity distribution of a gas at fixed temperature).

Solution The result $\langle x \rangle_N = 0$ is obtained imposing $N_{5,6} = N_{1,2} := K$, while $N_{3,4} = N - 2K$. In other words:

$$N(x = 1) = N(x = -1) = K; \qquad N(x = 0) = N - 2K. \tag{5.58}$$

The analogy with the Maxwell-Boltzmann gas suggests us that the probability $P(x|\langle x \rangle_N = 0)$, in the large N limit, is obtained as the most likely value of the ratio $N(x)/N$:

$$P(x|\langle x \rangle_N = 0) = \lim_{N \to \infty} \frac{\bar{N}(x)}{N}. \tag{5.59}$$

To determine $\bar{N}(x)$, we must find the maximum of the probability $P(K)$, that is just the multinomial distribution

$$P(K) = \frac{N!}{(K!)^2(N - 2K)!} \frac{1}{3^K} \frac{1}{2^{N-2K}} \frac{1}{6^K}.$$

We obtain, using the Stirling approximation $\ln N! \simeq N \ln N - N$:

$$\ln P(K) \simeq N \ln N - 2K \ln K - (N - 2K)\ln(N - 2K) - K \ln 18 - (N - 2K)\ln 2,$$

of which we must determine the maximum with respect to K. Carrying out the algebra, we obtain

$$\frac{\bar{K}}{N} = \left(2 + \frac{3}{\sqrt{2}}\right)^{-1},$$

which, substituted into Eqs. (5.58) and (5.59), gives us the required result.

Problems 3 Consider an ideal monoatomic gas. The energy of each atom has an internal quantized component

$$\epsilon = \frac{p^2}{2m} + n\Delta\epsilon, \qquad n = 0, 1, 2, \ldots, \qquad \Delta\epsilon > 0.$$

- Determine the Helmholtz free energy of the system.
- Determine the heat capacity C_V of the system.

5.9 Further Reading

The basic reference for the material in this section is:

- L.D. Landau, E.M. Lifsits, *Statistical Physics* (Pergamon Press, 1969)

On issues concerning ergodicity and mixing:

- A.I. Khinchin, *Mathematical Foundations of Statistical Mechanics* (Dover, 1949)
- H.G. Schuster and W. Just, *Deterministic chaos*. An introduction. *Fourth edition* (Wiley, 2005. Available online)

  For in-depth coverage of magnetic systems and critical phenomena, see e.g.:

- M. Pischke and B. Bergersen, *Equilibrium Statistical Physics* (World Scientific, 2006)

Appendix

A.1 Perturbation Methods

The calculation of the partition function for an ideal gas, that we carried out in Sect. 5.4.1, ended up consisting of an evaluation of Gaussian integrals. In the presence of interactions, however, the Hamiltonian $\mathcal{H}(\Gamma)$ will contain non-quadratic terms that make the evaluation of the integrals in Eq. (5.29) not possible, in general, in closed form. Approximation methods become then necessary.

Two possibilities that become immediately apparent, from inspection of Eq. (5.29), are high temperature and low temperature expension of the integrand. The first case corresponds to a regular expansion in the interaction term of the Hamiltonian, using the factor $1/T$ as expansion parameter. The second case corresponds to a saddle point (Laplace) kind of expansion of the integral. We illustrate the two procedures on two specific examples.

A.1.1 High Temperature Expansion

We consider the case of real gas, in which molecules interact with a two-body potential:

$$\mathcal{H} = \frac{1}{2m} \sum_i p_i^2 + \sum_{i \neq j} U(\mathbf{q}_i - \mathbf{q}_j). \tag{5.60}$$

The expansion of Eq. (5.29) is carried out with respect to the interaction term U:

$$Z = \frac{1}{N! \delta \Gamma} \int d^{3N} p \, d^{3N} q \, \exp\left(-\frac{1}{2mT} \sum_i p_i^2\right)\left[1 - \frac{1}{T} \sum_{i \neq j} U(\mathbf{q}_i - \mathbf{q}_j)\right.$$
$$\left. + \frac{1}{2T^2} \sum_{i \neq j} \sum_{k \neq l} U(\mathbf{q}_i - \mathbf{q}_j) U(\mathbf{q}_k - \mathbf{q}_l) + \dots \right]. \tag{5.61}$$

Noticing that the kinetic part decouples from the interaction part in the integral, Eq. (5.61), can be rewritten in the form

$$Z = Z_0\left[1 - \frac{N(N-1)}{TV^2} \int d^3 q_1 d^3 q_2 \, U(\mathbf{q}_1 - \mathbf{q}_2) + \dots \right],$$
$$= Z_0\left[1 - \frac{N}{Tv} \int d^3 q \, U(\mathbf{q}) + \dots \right], \tag{5.62}$$

where $v = V/N$ is the specific volume of one molecule and

$$Z_0 = \frac{V^N}{N! \delta \Gamma} \int d^{3N} p \, \exp\left(-\frac{1}{2mT} \sum_i p_i^2\right) \tag{5.63}$$

is the expression of the partition function in the ideal case, already provided in Eq. (5.31).

We call attention to the structure of the expansion in Eq. (5.61), that appears to be, essentially, an expansion in the interaction length of the potential. To lowest order, $Z = Z_0$, the particles become totally independent, as the effect of interaction is killed by thermal motion. To next order, we have a linear correction, that is the contribution from the average potential energy of the molecules, in the field of the other molecules in the gas. No molecule correlations are taken into account at this level. Proceeding in the perturbation expansion, clusters involving larger number of particles must be taken into account, hence the name for this perturbative approach, of **cluster expansion**. These increasingly more complex clusters account for the way in which higher order correlations contribute to the dynamics, as the interaction becomes stronger.

A.1.2 Low Temperature Expansion

For $T \to 0$, the system will tend to lie at the lowest possible energy levels available to the dynamics. As temperature decreases, also the fluctuations around the zero energy will become smaller. This again simplifies life in the case the Hamiltonian can be expressed as a quadratic free component, plus a higher order interaction term.

Let us consider for simplicity the case of a one dimensional enharmonic oscillator:

$$\mathcal{H} = \frac{p^2}{2m} + \frac{\alpha q^2}{2} + \lambda q^4, \tag{5.64}$$

with α and λ positive to make the dynamics finite, and have a single potential minimum at $q = 0$. The Hamiltonian in Eq. (5.64) could model e.g. a mesoscopic system in contact with a thermal bath, for which fluctuations are small but not negligible. The thermodynamics of the system will be described again by the partition function

$$Z = (\delta\Gamma)^{-1} \int dpdq \, \exp\left(-\frac{1}{T}\left(\frac{p^2}{2m} + \frac{\alpha q^2}{2} + \lambda q^4\right)\right) \tag{5.65}$$

The difficult part of the evaluation of Z lies in the quartic term λq^4, that makes the integral non-Gaussian. Nevertheless, for $T \to 0$, we expect fluctuations to be smaller, and the quartic term to become a correction. This condition can be made explicit, rescaling the coordinate part of the partition function

$$q \to \hat{q} = (\alpha/T)^{1/2}q.$$

Substituting into Eq. (5.65), after carrying out the integral over the momentum:

$$Z = \frac{(\pi m)^{1/2}T}{\alpha^{1/2}\delta\Gamma} \int d\hat{q} \, \exp\left(-\frac{\hat{q}^2}{2} + \hat{\lambda}\hat{q}^4\right), \quad \hat{\lambda} = \frac{T^2\lambda}{\alpha^2}. \tag{5.66}$$

We thus see that the low temperature expansion converts into an expansion in the (rescaled) coupling constant $\hat{\lambda}$:

$$Z = \frac{(\pi m)^{1/2}T}{\alpha^{1/2}\delta\Gamma} \int d\hat{q} \, \exp\left(-\frac{\hat{q}^2}{2}\right)\left[1 + \hat{\lambda}\hat{q}^4 + \frac{1}{2}\hat{\lambda}^2\hat{q}^8 + \cdots\right]. \tag{5.67}$$

Chapter 6
The Theory of Fluctuations

6.1 Fluctuations of Thermodynamic Quantities

Thermodynamic systems at equilibrium are characterized by fluctuations that are a manifestation of their discrete nature at microscopic scales. We have already seen some examples of such fluctuations: those in the number of particle ΔN in a certain volume V of a gas; the energy fluctuations in a system in contact with a heat bath. In the first case, we found the typical result of a Poisson process [see Eq. (2.29)]:

$$\Delta N \sim \bar{n} V.$$

In the second, the fluctuation distribution was governed by the canonical distribution Eq. (5.27), which gave for the fluctuation amplitude:

$$\Delta E \sim T.$$

In all cases, it is clear that an important role in the determination of the fluctuation amplitude, must be played by the Boltzmann entropy of the fluctuation macrostates. We recall that the probability of a generic macrostate $\mathbf{m}$ of an isolated system, is proportional to the number of microstates $\Delta \hat{\Gamma}(\mathbf{m})$, with which the macrostate can be realized. Exploiting Eq. (5.12), this number can be expressed in terms of the Boltzmann entropy $S(\mathbf{m})$:

$$S(\mathbf{m}) = \exp(\Delta \hat{\Gamma}(\mathbf{m})). \tag{6.1}$$

A PDF for the macrostates $\mathbf{m}$ can then be obtained, exploiting Eq. (5.10) and the definition of microcanonical distribution, Eq. (5.7):

$$\rho(\mathbf{m}) = \rho^M(\Gamma(\mathbf{m})) \frac{\Delta \Gamma(\mathbf{m})}{\Delta \mu(\mathbf{m})} = \frac{\Delta \Gamma(\mathbf{m})}{\Delta \mu(\mathbf{m}) \Delta \Gamma(E)},$$

© Springer International Publishing Switzerland 2015
P. Olla, *An Introduction to Thermodynamics and Statistical Physics*,
UNITEXT for Physics, DOI 10.1007/978-3-319-06188-7_6

where $\Gamma(\mathbf{m})$ is the generic point in $\Delta\Gamma(\mathbf{m})$, and $\Delta\mu(\mathbf{m})$ is the μ-space element corresponding to $\mathbf{m}$. Exploiting Eq. (6.1), together with the fact that $\Delta\mu(\mathbf{m})$ and $\Delta\Gamma(E)$ are constant, we find trivially

$$\rho(\mathbf{m}) = C \exp(S(\mathbf{m})). \qquad (6.2)$$

From here, we can determine the distribution of the fluctuation $z_k = m_k - \bar{m}_k$ of the generic thermodynamic quantity m_k. The function $S(\mathbf{m})$ varies on the scale of the equilibrium value $\bar{\mathbf{m}}$, hence, for variations of $\mathbf{m}$ on the scale of the thermal fluctuations, we can Taylor expand:

$$S(\mathbf{m}) \simeq \bar{S} - \frac{1}{2} \sum_{ij} \alpha_{ij} z_i z_j, \qquad (6.3)$$

where $\bar{S} \equiv S(\bar{\mathbf{m}})$ is the equilibrium (Gibbs) entropy. We thus see that the fluctuation PDF is a Gaussian, with correlation matrix $\langle z_i z_j \rangle = (\alpha^{-1})_{ij}$.

Note Let us verify that $(\alpha^{-1})_{ij}$ is the correlation matrix. We notice at once that the antisymmetric part of α_{ij} does not contribute in the sum $\alpha_{ij} z_i z_j$, and it can identically be put equal to zero. The matrix α_{ij} can be diagonalized by means of a unitary transformation. The Jacobian of the corresponding change of variables in the PDF, is therefore equal to one, and we have

$$\rho_{\hat{\mathbf{z}}}(\hat{\mathbf{z}}) = \rho_{\mathbf{z}}(\mathbf{z}(\hat{\mathbf{z}})) = C \exp(-\frac{1}{2} \sum_k \hat{\alpha}_{kk} \hat{z}_k^2),$$

where the hat indicates variables in the diagonal representation. But $(\hat{\alpha}^{-1})_{ik} \equiv \hat{\alpha}_{kk}^{-1} \delta_{ik} = \langle \hat{z}_i \hat{z}_k \rangle$ defines the correlation matrix in the diagonal representation, while $(\alpha^{-1})_{ik}$ is its form in the original representation.∎

We must at this point determine the coefficients α_{ij}, which is equivalent to ask what is the form of the function $S(\mathbf{m})$. This function could be determined from the external work ΔW required to generate the fluctuation $\mathbf{z} = \mathbf{m} - \bar{\mathbf{m}}$, through the formula $\Delta S = -\Delta W/T$.

To make the concept more precise, we follow the approach in Sect. 4.7, and consider a portion of the isolated system, that can exchange work and heat with the rest of the system, but whose number of molecules remains constant. We are interested in the fluctuations of the thermodynamic quantities that define the state of this portion of the system. Let us focus on the volume V and the entropy S (as usual, quantities with no subscripts will refer to the subsystem, while subscript *tot* will identify quantities referred to the isolated system).

In correspondence to the fluctuations of S and V, there will be a departure from equilibrium of the whole isolated system, and therefore a decrease of its entropy, $\Delta S_{tot} < 0$. This decrease in the total entropy, corresponds to a departure from

Fig. 6.1 The same entropy variation could be seen, both as the result of a fluctuation, and (with opposite sign), of the relaxation to equilibrium, after the system has been pushed out of equilibrium by an external action (the work ΔW^{min})

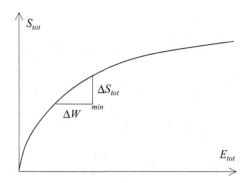

the equilibrium curve $\bar{S}_{tot} = \bar{S}_{tot}(E_{tot})$, that could be obtained in alternative way, executing work from the outside, starting from a lower energy condition (see Fig. 6.1).

Let us go back for a moment to the issue of the minimum work required to generate the fluctuation. Recall the result of Eq. (4.51), valid in the case two parts of the system were initially in equilibrium:

$$\Delta W^{min} = \Delta E - T\Delta S + P\Delta V. \tag{6.4}$$

The modification was produced in such a way that the total volume of the isolated system did not change (an example of such action could be a paddle that stirs the gas in a fixed volume tank). Thus ΔW^{min} is the increase of energy of the whole isolated system.

Now, the modification associated with ΔV and ΔS, will generate a small variation in the pressure and the temperature of the subsystem, that will be pushed slightly out of equilibrium. After a while, equilibrium is recovered, and the final state of the system will be the same as if an amount of heat equal to the work ΔW^{min}, had been provided. In response to this, the entropy of the system will increase by an amount $\Delta S_{tot} = \Delta W^{min}/T$. Hence, $\Delta W^{min}/T$ is the entropy growth in the isolated system, in the relaxation to equilibrium from the initial non-equilibrium condition.

Summarizing, the probability of a fluctuation associated with variations ΔV and ΔS in the subsystem, will be given by, utilizing Eq. (6.4):

$$\rho(\Delta V, \Delta S) = C \exp(-\Delta W^{min}/T) = C \exp\left(-\frac{\Delta E - T\Delta S + P\Delta V}{T}\right). \tag{6.5}$$

We notice at once that, since the modifications take place at equilibrium, the temperature and the pressure, inside and outside the subsystem, must be equal. The work ΔW^{min}, that is necessary to exert to produce the fluctuation, comes therefore from the departure from equilibrium of the internal pressure and temperature of the subsystem. This tells us that the minimal work must be at least quadratic in ΔV and ΔS. In fact, things must be this way, if the transformation in the isolated system is carried out around an equilibrium configuration corresponding to a maximum of S^{tot}. This is confirmed by Eq. (6.4), whose RHS is zero at linear order, simply by the first law of thermodynamics.

We must expand the energy of the system to second order in ΔV and ΔS:

$$\Delta E \simeq T \Delta S - P \Delta V + \frac{\partial^2 E}{\partial S^2} (\Delta S)^2 + 2 \frac{\partial^2 E}{\partial S \partial V} \Delta S \Delta V + \frac{\partial^2 E}{\partial V^2} (\Delta V)^2.$$

Substituting into Eq. (6.4), we find

$$\Delta W^{min} \simeq \frac{\partial^2 E}{\partial S^2} (\Delta S)^2 + 2 \frac{\partial^2 E}{\partial S \partial V} \Delta S \Delta V + \frac{\partial^2 E}{\partial V^2} (\Delta V)^2.$$

On the other hand, we can write

$$\frac{\partial^2 E}{\partial S^2} (\Delta S)^2 = \frac{\partial T}{\partial S} (\Delta S)^2 = \left(\Delta T - \frac{\partial T}{\partial V} \Delta V \right) \Delta S,$$

$$\frac{\partial^2 E}{\partial V^2} (\Delta V)^2 = -\frac{\partial P}{\partial V} (\Delta V)^2 = -\left(\Delta P - \frac{\partial P}{\partial S} \Delta S \right) \Delta V,$$

$$\frac{\partial^2 E}{\partial S \partial V} = \frac{\partial T}{\partial V} = -\frac{\partial P}{\partial S},$$

and therefore, substituting again into ΔW^{min}, and then into $\rho(\Delta V, \Delta S)$ (see Eq. (6.5)):

$$\rho(\Delta V, \Delta S) = C \exp\left(\frac{\Delta P \Delta V - \Delta T \Delta S}{T} \right), \qquad (6.6)$$

where ΔP and ΔT are considered here as functions of ΔS and ΔV. The above expression underlines the fact that the deviation from equilibrium must be associated with reaction forces, that are precisely the variations ΔT and ΔP (compare with the discussion in Sect. 4.2.2).

We notice at this point that we have four fluctuating quantities, that are functions of only two independent parameters. The parameters that are chosen as independent, are those of which the fluctuation amplitude can be calculated explicitly. We provide an example of such calculation in the case of the pair T, V.

We have in this case to express the variations ΔS and ΔP, in Eq. (6.6), as functions of ΔT and ΔV. We have:

$$\Delta S = \frac{\partial S}{\partial T} \Delta T + \frac{\partial S}{\partial V} \Delta V = \left(\frac{\partial S}{\partial E} \right)_V \left(\frac{\partial E}{\partial T} \right)_V \Delta T - \frac{\partial^2 F}{\partial V \partial T} \Delta V$$

$$= \frac{C_V}{T} \Delta T - \frac{\partial^2 F}{\partial V \partial T} \Delta V$$

and

$$\Delta P = \left(\frac{\partial P}{\partial T} \right)_V \Delta T + \left(\frac{\partial P}{\partial V} \right)_T \Delta V = -\frac{\partial^2 F}{\partial V \partial T} \Delta T + \left(\frac{\partial P}{\partial V} \right)_T \Delta V.$$

Substituting into Eq. (6.2):

$$\rho(\Delta V, \Delta T) = C \exp\left(-\frac{C_v}{2T^2}(\Delta T)^2 + \frac{1}{2T}\left(\frac{\partial P}{\partial V}\right)_T (\Delta V)^2\right), \tag{6.7}$$

from which we find:

$$\langle(\Delta T)^2\rangle = \frac{T^2}{C_v}, \quad \langle\Delta T \Delta V\rangle = 0; \quad \langle(\Delta V)^2\rangle = -T\left(\frac{\partial V}{\partial P}\right)_T. \tag{6.8}$$

The same procedure leading to Eq. (6.8) could be utilized with a different choice of independent variables, and, in this way, it is possible to determine the fluctuation amplitude of the other thermodynamic parameters of the subsystem.

6.2 Spatial Structure of Fluctuations

As we have seen repeatedly. the fluctuation of an extensive quantity (e.g. the number of particle), referred to a portion of a system of volume V, will scale as $V^{1/2}$, while the corresponding mean value will scale as V. Similarly, the fluctuations of an intensive quantity (think of the density) will scale as $V^{-1/2}$, while the corresponding mean quantity will be a constant. In both cases, the relative size of fluctuation and mean quantity will be $V^{-1/2}$. More general behaviors could be expected far from equilibrium, or in critical conditions.

It is rather clear that the spatial structure of the fluctuations of an extensive quantity could be determined from that of the associated density-like intensive quantity. It is thus to the second, that we direct now our attention. As already seen in Sects. 2.3 and 3.3, the fluctuation properties of an intensive quantity (in that case, the density) are those of a random field. This suggests us to resort to the techniques of correlation analysis and coarse-graining, introduced in Sects. 2.5 and 2.6.

Let us indicate with $\phi(\mathbf{x}, t)$ the fluctuation component of the field, and introduce its coarse grained version

$$\phi_V(\mathbf{x}, t) = \langle\phi(\mathbf{x}, t)\rangle_V. \tag{6.9}$$

Information on the spatial structure of the fluctuations can be retrieved from the scale dependence of the fluctuation amplitude $\langle\phi_V^2\rangle$. In alternative, we could analyze the behavior of the correlation function

$$C(\mathbf{x}, \mathbf{x}') = \langle\phi(\mathbf{x}', t)\phi(\mathbf{x}, t)\rangle.$$

The two quantities $C(\mathbf{x}, \mathbf{x}')$ and $\langle\phi_V^2\rangle$ are not independent. We consider the case in which $V^{1/3}$ is much smaller than the inhomogeneity scale of the system, so that the the statistics can be considered spatially uniform, and assume isotropy, so that

$$\langle\phi(\mathbf{x}', t)\phi(\mathbf{x}, t)\rangle = C(|\mathbf{x} - \mathbf{x}'|).$$

In this case, the fluctuation amplitude for ϕ_V can be written in the form

$$\langle \phi_V^2 \rangle = \frac{1}{V^2} \int_V d^3x \int_V d^3x' \langle \phi(\mathbf{x}', t)\phi(\mathbf{x}, t) \rangle$$

$$\sim \frac{1}{V} \int_V d^3x\, C(x). \tag{6.10}$$

We analyze the behavior of Eq. (6.10) separately in the two limits $V^{1/3} \ll \lambda_\phi$ and $V^{1/3} \gg \lambda_\phi$, where λ_ϕ is the correlation length for the fluctuations.

The situation is illustrated in Fig. 6.2. The behavior for $V^{1/3} \ll \lambda_\phi$ reflects that for small separations of $C(x)$, which, as we have seen in the case of the density fluctuations (see Eq. (3.13)), is usually going to be singular. We shall analyze the small separation range more in detail in Sect. 6.2.1. For $V^{1/3} \gg \lambda_\phi$, we find instead:

$$\langle \phi_V^2 \rangle \sim V^{-1}, \tag{6.11}$$

independent of the profile of $C(x)$. This is the behavior expected in the independent molecule picture of an ideal gas, already observed in Eq. (3.12).

To give a more precise dynamical content to Eqs. (6.10) and (6.11), let us consider, as a concrete example, the fluctuations of the specific volume v:

$$\phi(\mathbf{x}, t) = \tilde{v}(\mathbf{x}, t) - \bar{v}. \tag{6.12}$$

We can then use the relation

$$\Delta V = V - \bar{V} \simeq V\phi_V, \tag{6.13}$$

which allows us to make contact with Eq. (6.8). Substituting Eqs. (6.12) and (6.13) into Eq. (6.8), we find:

$$\langle \phi_V^2 \rangle = -\frac{T}{N}\left(\frac{\partial v}{\partial P}\right)_T. \tag{6.14}$$

Fig. 6.2 Profile of the fluctuation amplitude $\langle \phi_V^2 \rangle$, in the case $\langle \phi^2 \rangle < \infty$

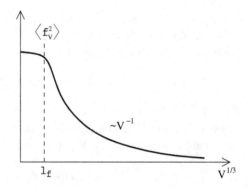

Of course, as we have already pointed out, it is impossible for Eq. (6.14) to produce a scaling for $\langle \phi_V^2 \rangle$ that is different from V^{-1}, as the only extensive quantity in the RHS of the equation is the factor N^{-1}.

What, in the derivation in Sect. 6.1, prevented us to obtain something different from such kind of behavior? We quickly realize that the problem lies in the way the energies and the entropies of the different parts in which the system was subdivided, added up to the total energy and entropy of the system, as if no correlation among these parts existed. In order to obtain a behavior for $\langle \phi_V^2 \rangle$, that differs from that in Eq. (6.11), such correlations should be taken into account from the start. But, as already mentioned several times, the correlations of microscopic fluctuations are themselves necessarily microscopic.

6.2.1 The Gaussian Model

In principle, the spatial structure of macroscopic fluctuations could be obtained directly from the distribution in Γ-space of the thermodynamic system, exactly as it was done in the microscopic case. Carrying out the necessary coarse-graining operations in the presence of macroscopic correlations, however, turns out to be often far from trivial.

The alternative approach is to try to determine the form of the fluctuation PDF, based on purely macroscopic considerations. Especially in the case of critical phenomena, this turns out to be a winning strategy, as the macroscopic fluctuations appear to be remarkably unsensitive on the details of the microscopic interactions: a phenomenon called **universality**.

We proceed to construction of the fluctuation PDF in the case we have just one fluctuating field $\phi(\mathbf{x}, t)$. The coarse graining scale of the field, Δx, is supposed much smaller than the correlation length λ_ϕ, but still macroscopic. As we are interested, for the moment, only in the spatial statistics of the field, and we assume stationary statistics, we can omit the dependence of ϕ on t. The field $\phi(\mathbf{x})$ will thus indicate a snapshot of the fluctuations at an arbitrary time.

As discussed in Sect. 2.5, the statistics of the random field $\phi(\mathbf{x})$ is contained in the functional PDF $\rho[\phi]$, that we suppose can be written in the form

$$\rho[\phi] = \frac{1}{Z} \exp(-\frac{1}{T} \mathcal{H}[\phi]). \tag{6.15}$$

(Square brackets indicate, as usual, dependence on the full field configuration; see Eq. (2.33)). We have written intentionally the PDF in a form that reminds the canonical distribution, with $\mathcal{H}$ the fluctuation Hamiltonian.

From the point of view of fluctuations, thermodynamic systems can be grouped roughly into three classes:

- Systems whose microscopic components interact by means of long-range forces. Long-range correlations are produced rather trivially by the fact that interactions between separate macroscopic parts of the system cannot be considered negligible.
- Systems whose microscopic parts interact with short-range forces, but are still characterized by long-range correlations. As discussed in Sect. 4.11.2, this is what defines a critical regime for the system. In this case, macroscopic correlations are the result of a cooperative behavior of the microscopic components.
- Systems whose microscopic parts interact with short-range forces, and are not critical, that are characterized by fluctuation with a correlation length that is essentially zero.

The interpretation of the functional $\mathcal{H}[\phi]$ appearing in Eq. (6.15) as a fluctuation energy, allows to distinguish the two situations in which long-range correlations arise from long-range interactions and cooperative effects.

If the interaction is short range, the energy variation generated by a perturbation on a microscopic component of the system, will not be affected by the state of the system, at macroscopic separation from the site of the perturbation. This means that the energy $\mathcal{H}[\phi]$ must be a sum of contributions from the different macroscopic regions of the system. In formulae:

$$\mathcal{H}[\phi] = \int\limits_{V_{tot}} \mathrm{d}^3x \, H(\phi(\mathbf{x}), \nabla\phi(\mathbf{x}), \nabla^2\phi(\mathbf{x}), \ldots), \qquad (6.16)$$

where H represents a local energy density of the fluctuations, and the '...' indicate higher order derivatives of the field in $\mathbf{x}$. If the interaction is long-range, the energy density in Eq. (6.16) will be non-local, and will depend on values of ϕ and its derivatives, also at points different from $\mathbf{x}$.

The PDF $\rho[\phi]$ in Eq. (6.15) will be in general non-Gaussian. The fact is that fluctuations at the macroscopic scale λ_ϕ, cannot in general be assumed to have small amplitude. This differs from the situation considered in Sect. 6.1, in which smallness of the fluctuations, and applicability of Eq. (6.3), was consequence of the fact that a thermodynamic limit could be assumed for any macroscopic portion V of the system.

Note Gaussian statistics could be recovered if a second coarse graining of the field, $\phi \to \phi_V$, is carried out at a scale $V^{1/3} \gg \lambda_\phi$. The coarse-grained field ϕ_V could in this case be seen as a sum of i.i.d. contributions from $\phi(\mathbf{x})$ in regions of size λ_ϕ. If $V^{1/3} \gg \lambda_\phi$, the number of contribution would become large, and central limit arguments would guarantee Gaussian statistics.∎

The non-Gaussian part of the statistics is typically responsible for the most interesting aspects of the system dynamics. This is especially true in the case of critical phenomena. Some general information on the spatial structure of fluctuations can nevertheless be obtained already in the case of a purely Gaussian statistics. We focus on the case of a local microscopic dynamics, and consider the simplified model statistics described by the quadratic form

$$\mathscr{H}[\phi] = \frac{1}{2} \int_{V_{tot}} d^3x \, \phi(\mathbf{x})[a - b\nabla^2]\phi(\mathbf{x}). \tag{6.17}$$

To obtain the statistics of ϕ, it is simpler to work in Fourier space. The same information on the field $\phi(\mathbf{x})$ is in fact contained in the Fourier components

$$\phi_{\mathbf{k}} = \int_{V_{tot}} d^3x \, \phi(\mathbf{x})e^{-i\mathbf{k}\cdot\mathbf{x}}.$$

In terms of Fourier transforms, the energy functional $\mathscr{H}[\phi]$ becomes

$$\mathscr{H}[\phi] = \frac{1}{2} \int \frac{d^3k}{(2\pi)^3} [a + bk^2]|\phi_{\mathbf{k}}|^2 = \frac{1}{2V_{tot}} \sum_{\mathbf{n}} [a + bk_{\mathbf{n}}^2]|\phi_{\mathbf{k_n}}|^2, \tag{6.18}$$

where the last term is the result of the discretization of Fourier space, $\Delta k = 2\pi V_{tot}^{-1/3}$, induced by the finiteness of the volume of the system. The same change of variable $\phi(\mathbf{x}) \to \phi_{\mathbf{k}}$ could be carried out with the PDF $\rho[\phi]$. Now, the operation of Fourier transform is linear, hence, the Jacobian in the change of variables is a constant. This means that the PDF $\rho[\phi]$ can be expressed in terms of Fourier components, by simply substituting Eq. (6.18) into Eq. (6.15). The result is

$$\rho[\phi] = \frac{1}{Z'} \exp\left\{ -\frac{1}{2TV_{tot}} \sum_{\mathbf{n}} [a + bk_{\mathbf{n}}^2]|\phi_{\mathbf{k_n}}|^2 \right\}. \tag{6.19}$$

The fluctuation amplitude in Fourier space is

$$\langle \phi_{\mathbf{k_n}} \phi_{\mathbf{k_m}}^* \rangle = \frac{V_{tot}T}{a + bk^2} \delta_{\mathbf{nm}},$$

which, taking the infinite volume limit, becomes:

$$\langle \phi_{\mathbf{k}} \phi_{\mathbf{k'}} \rangle = \frac{T}{a + bk^2}(2\pi)^3 \delta(\mathbf{k} + \mathbf{k'}) := (2\pi)^3 C_{\mathbf{k}} \delta(\mathbf{k} + \mathbf{k'}). \tag{6.20}$$

We recognize in Eq. (6.20) the expression for the correlation of Fourier modes of Eq. (2.62), in which the energy spectrum $C_{\mathbf{k}}$ is the Fourier transform of the correlation $C(\mathbf{x}) = \langle \phi(\mathbf{x})\phi(0) \rangle$ (Wiener-Khinchin theorem). Carrying out the inverse Fourier transform, we obtain:

$$C(\mathbf{x}) = \frac{T}{\pi b} \frac{e^{-\sqrt{a/b}\,x}}{x}, \tag{6.21}$$

that tells us three things:

First, the value of the correlation length

$$\lambda_\phi \sim (b/a)^{1/2}. \tag{6.22}$$

Second, the fact that, at criticality, the correlation function takes a power-law form

$$C(\mathbf{x}) = \frac{T}{\pi b x}. \tag{6.23}$$

Third, the fact that the fluctuations have an important component at the discretization scale Δx. Equation (6.22) would lead in fact us to expect $\langle \phi^2 \rangle \sim \Delta x^{-1}$.

As already mentioned, divergent behaviors like this are not uncommon in random fields, and should not be surprising in an intensive density-like quantity such as ϕ. Such divergences are the counterpart, in dimension greater than one, of the non-differentiable behaviors observed in stochastic processes (see discussion at the end of Sect. 2.6). The simplest way to cure such divergence, is to coarse grain ϕ at an appropriate scale. In analogy with Eq. (2.51), we can write ϕ_V as a convolution with a function $G_V(\mathbf{x})$, with support in a volume V around $\mathbf{x} = 0$, with $\int d^3x\, G_V(\mathbf{x}) = 1$ and $G_V(\mathbf{x}) > 0$:

$$\phi_V(\mathbf{x}) = \int d^3y\, G_V(\mathbf{x} - \mathbf{y})\phi(\mathbf{y}).$$

Thus, in analogy with Eq. (2.66):

$$\phi_{V,\mathbf{k}} = G_{V,\mathbf{k}}\phi_{\mathbf{k}},$$

where $G_{V,\mathbf{k}} = 0$ for $k \gg V^{-1/3}$, while $G_{V,\mathbf{k}} = 1$ in the opposite limit. We can carry out with ϕ_V the same operation as in the case of ϕ. We have

$$\langle \phi_{V,\mathbf{k}}\phi_{V,\mathbf{k'}} \rangle = (2\pi)^3 C_{V,\mathbf{k}}\delta(\mathbf{k} + \mathbf{k'}),$$

and, at the same time

$$\langle \phi_{V,\mathbf{k}}\phi_{V,\mathbf{k'}} \rangle = |G_{V,\mathbf{k}}|^2 \langle \phi_{\mathbf{k}}\phi_{\mathbf{k'}} \rangle = (2\pi)^3 |G_{V,\mathbf{k}}|^2 C_{\mathbf{k}}\delta(\mathbf{k} + \mathbf{k'}),$$

i.e.

$$C_{V,\mathbf{k}} = |G_{V,\mathbf{k}}|^2 C_{\mathbf{k}} \approx \frac{T}{a + bk^2}\theta(1 - Vk^3), \tag{6.24}$$

where $\theta(x)$ is the Heaviside theta function: $\theta(x > 0) = 1$, $\theta(x < 0) = 0$. We can then express the fluctuating amplitude of the coarse-grained field, exploiting again the Wiener-Khinchin theorem, as the integral of the cut-off energy spectrum $C_{V,\mathbf{k}}$. We find, from Eq. (6.24):

$$\langle \phi_V^2 \rangle = \int \frac{d^3 k}{(2\pi)^3} \, C_{V,\mathbf{k}} \approx \frac{T}{2\pi^2} \int\limits_0^{V^{-1/3}} \frac{k^2 dk}{a + bk^2}. \tag{6.25}$$

The divergent behaviors of $C(\mathbf{x})$, at $x = 0$, is thus transferred to that of $\langle \phi_V^2 \rangle$ for $V \to 0$:

$$\langle \phi_V^2 \rangle \sim \frac{T}{bV^{1/3}}, \qquad V^{1/3} \ll \lambda_\phi, \tag{6.26}$$

but, as long as V is finite, the fluctuation amplitude remains finite. We recover the V^{-1} decay of $\langle \phi_V^2 \rangle$ at $V^{1/3} > \lambda_\phi$, as expected, with the factor a^{-1} playing, with respect to ϕ_V, the same role played in Eq. (6.8), with respect to $\Delta V/V$, by the compressibility $-V^{-1}(\partial V/\partial P)_T$. We have in fact from Eq. (6.25):

$$\langle \phi_V^2 \rangle \sim \frac{T}{aV}, \qquad V^{1/3} \gg \lambda_\phi. \tag{6.27}$$

In this range, the fluctuation energy $\mathcal{H}[\phi_V]$ takes the same role of the quantity W^{min} in Eq. (6.8), that is the minimum work to produce the fluctuation ϕ_V in the isolated system.

6.2.2 Scaling Relations at the Critical Point

As discussed in Sect. 4.11.2, the approach to a critical point is signaled by divergence of the susceptibility of some order parameter of the system, together with divergence of the correlation length of the order parameter fluctuations. We have seen that, in the Gaussian model, the role of susceptivity is played by the parameter a^{-1}. By means of the scaling relation, Eq. (4.83), the approach to the critical point is described by the relation

$$a \sim t^\gamma, \qquad t = \frac{T - T_c}{T_c}, \tag{6.28}$$

while Eq. (6.22) gives us

$$\lambda_\phi \sim b^{1/2} t^{-\gamma/2} \sim t^{-\nu}. \tag{6.29}$$

The scaling of the correlation length, and that of the susceptivity, could be expressed one as a function of the other, provided b is itself a non-scaling quantity. The parameter b enters as a prefactor in the expression of the correlation correlation function provided by Eq. (6.21), and in that for the fluctuation amplitude for the coarse grained field ϕ_V at $V^{1/3} \ll \lambda_\phi$ (see Eq. (6.26)). This fluctuation amplitude remains finite at the critical point, the only divergent quantity being the correlation length. In fact, the statement that b remains finite as $t \to 0$, could be converted into one on the fact that fluctuations in the scaling region $x \ll \lambda_\phi$, must be completely independent of properties of the system at scale λ_ϕ. In any case, what we find is that b must tend to a constant at $t = 0$. This leads us to the relation between scaling exponents

$$\nu = \gamma/2.$$

This relation is valid only for the simple Gaussian model we have considered so far. The analysis of more complicated non-Gaussian models leads typically to anomalous scaling of the correlation function. From dimensional analysis:

$$C(\mathbf{x}) = \frac{c\lambda_\phi^\eta}{b} \frac{e^{\sqrt{b/a}\,x}}{x^{1+\eta}} \tag{6.30}$$

where a^{-1} continues to be the system susceptibility associated with ϕ, and c is a dimensionless constant. If again we require that a finite limit for $C(\mathbf{x})$ exists for $t \to 0$, we must have

$$b \sim \lambda_\phi^\eta \sim t^{-\eta\nu},$$

which, together with Eqs. (6.28) and (6.29), gives us

$$2 - \eta = \frac{\gamma}{\nu}. \tag{6.31}$$

6.3 Time Structure of Fluctuations

Studying the temporal structure of fluctuations, leads us necessarily to consider the dynamical properties of the system. We have seen that, at least in the case of ideal gases, kinetic theory is able to describe such dynamical properties. We quickly realize, however, that this theory is unable to cope with fluctuations, being based on a description of the system in terms of averages (albeit with respect to a generic time-dependent non-equilibrium PDF). A possible approach could be to extend kinetic theory beyond mean field, deriving evolution equations for the two-particle distribution $f_2(\mathbf{x}_1, x_2; \mathbf{v}_1, v_2; t)$, and, from there, to obtain fluid equations that keep into account the effect of fluctuations. Unfortunately, the approach is in general rather cumbersome, and, for some reason, closure approximations at higher order in the BBGKY hierarchy, seem not to work as well as the mean field approximation.

The alternative strategy (followed by A. Einstein), is to suppose that the behavior of the "exact" fluid quantities $\tilde{n}(\mathbf{x}, t)$, $\tilde{\mathbf{u}}(\mathbf{x}, t)$, etc., differs little, at equilibrium, from that of the corresponding averages $n(\mathbf{x}, t)$, $\mathbf{u}(\mathbf{x}, t)$, etc. This implies that the fluctuating quantities will obey, within a correction that will depend on microscopic effects to be identified, the same fluid (or kinetic) equation that would be obtained linearizing the original equations for the corresponding mean quantities. For instance, the heat transport equation (3.57) for the fluctuations in a quiescent fluid, will now have the form

$$\frac{\partial \Delta T}{\partial t} + \frac{2}{3}T\nabla \cdot \Delta \mathbf{u} - \kappa \nabla^2 \Delta T = \xi,$$

where $\xi(\mathbf{x}, t)$ is the term that describes the microscopic effects responsible for fluctuations. (To obtain a closed system of equations, similar formulae should be introduced for the density and velocity fluctuations).

From exam of this particular example, we understand the situation: in the absence of the term ξ, the diffusivity (and the viscosity in the Navier-Stokes equation (3.56), through destruction of $\Delta \mathbf{u}$), would smooth out and destroy any fluctuating component ΔT. The term ξ acts therefore, in all respect, as a forcing. Furthermore, while diffusivity acts on the macroscopic time scale L^2/κ, where L is the scale of the fluctuations that we are considering, ξ will be a sum of contributions by the individual molecules, that cancel one another on the average. On macroscopic time scales, these contributions can also be considered as independent. Hence, the forcing term ξ can be considered to have zero average, and to be uncorrelated in time. A stochastic process with such characteristics is called a **white noise**.

The fluctuations of a macroscopic variable in a thermodynamic system, will thus be described, in general, by an equation with a random forcing and a linear dissipation. In the simple case of a single variable $z = m - \bar{m}$:

$$\dot{z}(t) + \gamma z(t) = \xi(t). \tag{6.32}$$

This is the kind of equation that we have to solve, to determine the time structure of the fluctuations of m. For situations that are far from equilibrium, the equation will become typically non-linear, but, even if its form could in general be impossible to determine, we know that it will have to describe a process of relaxation to the equilibrium $\bar{m}$. Superimposed to the relaxation process, there will be a fluctuating component, that, negligible in the initial phase, will represent the bulk of the variation of m at equilibrium. The Boltzmann entropy $S(m)$ will grow in the process, and, superimposed to it, there will be a fluctuating component that will be the only one that depends on time at equilibrium.

Most of the information on the temporal structure of the fluctuations, is contained in the correlation function $C(t) = \langle z(t)z(0)\rangle$, that we can express in terms of the conditional average $\langle z(t)|z(0)\rangle$, by means of the relation

$$C(t) = \langle \langle z(t)|z(0)\rangle z(0)\rangle.$$

The equation for $\langle z(t)|z(0)\rangle$ is obtained, for $t > 0$, taking the conditional average of Eq. (6.32):

$$(\frac{d}{dt} + \gamma)\langle z(t)|z(0)\rangle = \langle \xi(t)|z(0)\rangle = 0. \tag{6.33}$$

We have exploited here the fact that $\xi(t)$ is uncorrelated with $z(0)$, which, for causality, can depend on $\xi(t')$ only for $t' < 0$. Multiplying Eq. (6.33) by $z(0)$, and taking the average, we obtain in the end:

$$(\frac{\mathrm{d}}{\mathrm{d}t} + \gamma)C(t) = 0,$$

that has solution $C(t) = \langle z^2 \rangle \mathrm{e}^{-\gamma t}$. At thermal equilibrium, the statistics for z must be stationary, so that $C(t) = C(-t)$, and we can write, for t generic:

$$C(t) = \langle z^2 \rangle \mathrm{e}^{-\gamma |t|}. \tag{6.34}$$

We thus see that the temporal scale of fluctuation of the quantity m is fixed by the dissipation time γ^{-1}.

6.3.1 Brownian Motion

An important application (and in fact the starting point of the theory of fluctuations) is the problem of Brownian motion. It is the random walk that small particles in a fluid (say pollen in water) execute in response to collisions with molecules of the fluid. The problem is the determination of the diffusivity of the particles. The theory of Brownian motion by A. Einstein, opened the way to some of the first measurements of the Boltzmann constant K. For this reason, limited to this section, we return to units in which $K \neq 1$.

The starting point of the theory, is that the liquid and the particle in suspension can be seen as an isolated system in equilibrium. A parameter that describes the particle, is its kinetic energy $E = \frac{1}{2}Mu^2$, that is distributed with the canonical distribution. Integrating over all the microscopic degrees of freedom (the water molecules and the molecules in the particle), we are left with:

$$\rho(\mathbf{u}) = C \exp(-\frac{E}{KT}). \tag{6.35}$$

From Eq. (6.35), we obtain immediately

$$\langle u^2 \rangle \sim \frac{KT}{M}.$$

At the same time, however, the motion of the particle in the fluid must be described by an equation in the form

$$\dot{\mathbf{u}} + \gamma \mathbf{u} = \boldsymbol{\xi}, \tag{6.36}$$

(Langevin equation) where $M\gamma$ is the drag coefficient of the particle in the fluid. Dimensionally, we can write

$$\frac{\gamma}{M} \sim \mu v R^{-2},$$

where v is the kinematic viscosity of the fluid (in the case of water $v \simeq 0.01\mathrm{cm}^2/\mathrm{s}$) and R is the dimension of the particle. We find immediately that the mean velocity of the particle is zero. Substituting into the expression for the time correlation for $\mathbf{u}$, given by Eq. (6.34):

$$\langle u(0)u(t) \rangle = \langle u^2 \rangle e^{-\gamma|t|} \sim \frac{KT}{M} \exp(-C(v/R^2)|t|).$$

From this information, we can obtain the diffusion coefficient of the Brownian particle, that is the ratio of the variance of its displacement, and the time in which the displacement takes place. This calculation must be carried out for times that are much longer than the correlation time γ^{-1}, in such a way that the particle motion can be considered in all respects a random walk. Indicating with $\Delta x(t)$ the particle displacement in the time t, we have obviously, from $\langle \mathbf{u} \rangle = 0$, that also $\langle \Delta \mathbf{x} \rangle = 0$, so that

$$\langle (\Delta x(t))^2 \rangle = \int_0^t d\tau \int_0^t d\tau' \langle u(\tau)u(\tau') \rangle.$$

Putting $t \gg \gamma^{-1}$ (compare with Eq. (2.50)), we obtain finally:

$$\kappa_B = \frac{\langle (\Delta x(t))^2 \rangle}{t} \sim \int_{-\infty}^{\infty} d\tau \langle u(0)u(\tau) \rangle \sim \frac{\langle u^2 \rangle}{\gamma} \sim \frac{KTR^2}{Mv}. \tag{6.37}$$

We notice that this equation, starting from macroscopic quantities, all easily measurable, allows to determine a purely microscopic parameter such as the Boltzmann constant K.

At this point, some words must be said on the use of terms, such as "microscopic" and "macroscopic", referred to the Brownian particle. None of them is in fact correct, and the particle should be defined more properly as "mesoscopic". The scale of the Brownian particle is certainly well above that of molecules, and its dynamics is described by macroscopic parameters, such as viscosity and drag. Nevertheless, microscale effects are not negligible, since they determine the particle mobility. It could be worth at this point to give a look at the relation between the Boltzmann and Shannon entropy of the system. This is in fact an example of a situation in which the scale of definition of the macrostate—in this case the velocity of the Brownian particle—is necessarily smaller than the fluctuation amplitude: $\Delta u \ll \sigma_u$. The contribution of the macroscopic degrees of freedom to the Shannon entropy, therefore, is not negligible. In analogy with Eq. (5.11), we have:

$$S = -\langle \ln(\rho N! \delta \Gamma) \rangle = \sum_{\mathbf{u}} P(\mathbf{u})S(\mathbf{u}) - \sum_{\mathbf{u}} P(\mathbf{u}) \ln P(\mathbf{u}),$$

where

$$S(\mathbf{u}) = - \int_{\Delta\Gamma(\mathbf{u})} d\Gamma \rho(\mathbf{u}) \ln(\rho(\mathbf{u})N! \delta \Gamma) = \ln(\Delta\Gamma(\mathbf{u})/(N! \delta \Gamma))$$

is the Boltzmann entropy of the macrostate $\mathbf{u}$, and $- \sum_{\mathbf{u}} P(\mathbf{u}) \ln P(\mathbf{u})$ is the Shannon entropy of the macroscopic degrees of freedom of the particle.

We notice the important relation between Boltzmann entropy, the dissipation term and the fluctuation of the thermodynamic variable. In the case of the Brownian motion, the thermodynamic variable is $\mathbf{u}$; its increase leads to an increase of the particle kinetic energy, a "cooling" of the system, and therefore a decrease of the entropy $S(\mathbf{u})$ associated with the thermal degrees of freedom. This decrease is the negative exponent $-E/T = -Mu^2/(2T)$ in the Gibbs distribution, that, non surprisingly, is quadratic in the fluctuating variable $\mathbf{u}$ (the equilibrium value of the variable is obviously $\bar{\mathbf{u}} = 0$). The reaction force that pushes $\mathbf{u}$ back to its equilibrium value is the drag force $-\Gamma\mathbf{u}$.

We notice that also the gradient with which entropy decreases is linear in $\mathbf{u}$. It would thus be interesting to imagine that the reaction force and the entropy gradient $\nabla_{\mathbf{u}} S(\mathbf{u})$ be in some way connected. We shall see that this connection is established by the fact that the solutions of the equation $\dot{\mathbf{u}} + \gamma\mathbf{u} = \boldsymbol{\xi}$ must be distributed with the PDF $\rho(\mathbf{u}) = C \exp(-Mu^2/(2T))$.

6.3.2 Fluctuation–Dissipation Relations

Fluctuations at equilibrium can be seen as the result of the balance between the action of the molecular forces and that of the friction forces. The first acts as a source of fluctuations. The second as a restoring force. The balance between the two determines the fluctuation amplitude. The same quantity is determined by the entropy of the system, Eqs. (6.2) and (6.3). We can thus eliminate, out of the three unknowns: molecular force, friction force and fluctuation amplitude, the molecular force, remaining in this way with just one relation between friction force and fluctuation amplitude. We have seen this strategy at work in the case of the Brownian motion. We present here a slightly different approach, that avoids to call into play the molecular forces.

As we have done now several times, let us isolate in our system a macroscopic part, described by a macroscopic variable $m(\{p_i, q_i\})$, where $\{p_i, q_i\} \equiv \Gamma$ are the canonical coordinates of the part in exam. Let us suppose, for simplicity, that thermal equilibrium corresponds to $\bar{m} = 0$.

Let us suppose now that we act on the subsystem with an external force f.

We see at once that there are basically two kinds of variables, that could be influenced by the external force. The first example is of the type of the Brownian particle velocity **u**: an external force, to keep the particle in motion, must balance the friction force $-\gamma\mathbf{u}$, thus generating an average power $\gamma\langle u^2\rangle$ in the system. The second example is provided by the coordinate **x** of the same Brownian particle, supposed trapped in a potential well (without potential well, there would not be an equilibrium value for **x**). In this case, the external force must balance the conservative force of the potential well; on the average, no power will be generated in the system.

The two situations are qualitatively different. In the dissipative case, the friction force is the reaction of the rest of the system to the macroscopic motion of the subsystem (think of the Brownian particle). In the conservative case, the reaction force is internal to the subsystem (think of a Brownian particle connected by a spring to a solid support; in this case the subsystem would include the Brownian particle as well as the spring and the support). In this second case, the dissipative forces by the rest of the system are still present, but will not act (on the average) if the external force is constant.

If the external force f is sufficiently small, we can expect the response of the subsystem to be linear:

$$\langle m\rangle_f \simeq \gamma_m f,$$

where $\langle .\rangle_f$ indicates average on the system in the presence of the force f. The **linear response** of the system is defined therefore by the relation

$$\gamma_m = \frac{d\langle m\rangle_f}{df}\bigg|_{f=0},\tag{6.38}$$

the quantity $d\langle m\rangle_f/df_{f=0}$ is called the **response function** of the subsystem.

Let us consider first the conservative case. Suppose that the subsystem could be separated from the rest of the system, and considered itself as an isolated system. Indicate with $E(\Gamma)$ the energy of the isolated subsystem, that we assume to be minimum for $m \equiv x = 0$. The external force required to maintain the subsystem in the new state will be therefore $f = -\partial E/\partial x$. A system with Hamiltonian $E(\Gamma) - fx(\Gamma))$ will have therefore equilibrium in $x(\Gamma) = x_f \neq 0$.

Let us consider now the subsystem as a part of the greater isolated system. The rest of the system will act as a thermal bath, and the subsystem will be described by the canonical distribution, Eqs. (5.27) and (5.30):

$$\rho(\hat{\Gamma}) = \exp\left(\frac{F(\langle x\rangle_f) - E(\Gamma) + fx(\Gamma)}{T}\right).\tag{6.39}$$

Now, x_f will be a fluctuating quantity, with average, from Eq. (6.39):

$$\langle x\rangle_f = \int d\hat{\Gamma}\, x(\Gamma)\exp\left(\frac{F(\langle x\rangle_f) - E(\Gamma) + fx(\Gamma)}{T}\right).\tag{6.40}$$

To obtain the response function, let us substitute Eq. (6.38) in Eq. (6.40), with $m = x$. We obtain, exploiting the condition $\partial_x F|_{x=0} = 0$ (the Helmholtz free energy is minimum at equilibrium):

$$\gamma_x = \frac{1}{T} \int d\hat{\Gamma} (x(\Gamma))^2 \exp\left(\frac{F(0) - E(\Gamma)}{T}\right) = \frac{\langle x^2 \rangle}{T}. \tag{6.41}$$

We have been able to determine the fluctuation amplitude for x, without having to deal with molecular forces. Equation (6.41) provides us the first example of **fluctuation–dissipation relation**, valid in the case of a dynamics dominated by conservative forces.

Note It is interesting to notice that, through derivatives of the partition function, in the presence of external forces, we can obtain averages of generic functions of the microscopic variables $\{p_i, q_i\}$. Equations (6.40) and (6.41) can be rewritten indeed in the form

$$\langle x \rangle_f = \frac{dZ(f)}{df} \quad e \quad \langle x^2 \rangle = \left.\frac{d^2 Z(f)}{df^2}\right|_{f=0},$$

that demonstrates the fact that the method of partition functions can be adapted to calculate correlation functions. This is not surprising, as $Z(f)$ is just the characteristic function (for imaginary value of the argument, see Sect. 2.1.3) of the random variable x. ■

Let us pass to the purely dissipative case. In this case, the variable of interest m is the "velocity" $u = \dot{x}$. No restoring forces will influence this variable, if the subsystem is separated by the rest of the system. The modified Hamiltonian, $E_f = E - fx$, will now describe a system in which the force f produces a constant acceleration. In the case of the Brownian particle, in fact, $\dot{p}_i = -\partial_{q_i} E_f = f/N$, and therefore, $u(t) = \dot{x}(t) = u(0) + ft$. Equation (6.40) becomes in this case:

$$\langle u \rangle_f = \int d\hat{\Gamma} u(\Gamma) \exp\left(\frac{F(\langle u \rangle_f) - E(\Gamma) + fx(\Gamma)}{T}\right), \tag{6.42}$$

and Eq. (6.38) leads to the result, imposing $m = u$:

$$\begin{aligned}
\gamma_u &= \frac{1}{T} \int d\hat{\Gamma} u(\Gamma) x(\Gamma) \exp\left(\frac{F(0) - E(\Gamma)}{T}\right) \\
&= \frac{1}{T} \int d\hat{\Gamma}\, u(t; \Gamma) \int_{-\infty}^{t} dt'\, \langle u(t')|\Gamma, t\rangle \exp\left(\frac{F(0) - E(\Gamma)}{T}\right) \\
&= \frac{1}{T} \int_{-\infty}^{0} dt \langle u(t) u(0) \rangle. \tag{6.43}
\end{aligned}$$

This relation, known under the name of **Green-Kubo formula**, provides the counterpart, valid in the case of purely dissipative systems, of the fluctuation-dissipation relation Eq. (6.41). It is easy to see that Eq. (6.43), in the case of the Brownian motion, reduces to Eq. (6.37).

Note We can generalize the fluctuation-dissipation relation to time-dependent situations. To do this, we modify our point of view slightly, focusing on the response force to a certain external perturbation. For instance, instead of trying to determine the mean position $\langle x \rangle_f$ of a particle in a potential well, to an external force f, we ask ourselves what will be the value of the mean reaction force $\langle X \rangle_x$ that will act on us, if we fix the particle position at x. We focus on the purely conservative problem, so that x will be associated with a certain energy of the system, and not with a heat influx.

For small perturbations, we can linearize:

$$E \to E_x \simeq E + Xx,$$

where $X = dE/dx = X(\Gamma)$ is the reaction force to x, that now, contrary to x itself, will be a fluctuating quantity. Suppose that the perturbation $x(\tau)$ was turned on at time $\tau = 0$, and that the system was originally in equilibrium with a thermal bath at temperature T. At $\tau = 0$, the system was therefore described by the canonical distribution

$$\rho(\Gamma, 0) = \rho_0(\Gamma) = \frac{1}{Z}e^{-\beta E}, \quad \beta = 1/T.$$

The evolution at later times is obtained solving the Liouville equation (5.5), that we can rewrite in the form $\partial_t \rho + \{E_x, \rho\} = 0$, where $\{.,.\}$ indicate Poisson parentheses with respect to Γ: $\{f, g\} = \partial_p f \partial_q g - \partial_q f \partial_p g$. Exploiting smallness of the perturbation, we can expand ρ with respect to x: $\rho = \rho_0 + \rho_1 + \cdots$ To lowest order in the expansion, we have $\rho = \rho_0$, that is the equilibrium distribution, and is stationary solution of the Liouville equation $\partial_t \rho_0 = -\{E, \rho_0\} = 0$. At $O(x)$, we find instead:

$$\partial_t \rho_1 + \{E, \rho_1\} := \partial_t \rho_1 + \hat{L}_0 \rho_1 = -x\{X, \rho_0\},$$

where we have introduced the new operator $\hat{L}_0 \equiv \{E, .\}$, that is a shorthand for the Poisson parentheses with respect to E. We can solve formally, imposing the initial condition $\rho_1(\Gamma, 0) = 0$ (equilibrium initial condition):

$$\rho_1(t) = -\int_0^t d\tau \, x(\tau)e^{-\hat{L}_0(t-\tau)}\{X, \rho_0\} = \beta \int_0^t d\tau \, x(\tau)e^{-\hat{L}_0(t-\tau)}\rho_0\{X, E\}$$

$$= -\beta \int_0^t d\tau \, x(\tau)e^{-\hat{L}_0(t-\tau)}\rho_0 \dot{X}, \tag{6.44}$$

where $\dot{X} = \{E, X\} \equiv \dot{\Gamma}\partial_\Gamma X(\Gamma)$ is the variation rate of $X(\Gamma)$, induced by the Hamiltonian flux of $E(\Gamma)$. From here we find, for the average response force $\langle X(t) \rangle_x \simeq \int d\Gamma \rho_1(\Gamma, t)X(\Gamma)$:

$$\langle X(t)\rangle_x = -\beta \int\limits_0^t d\tau \, x(\tau) \int d\Gamma \, X e^{-\hat{L}_0(t-\tau)} \rho_0 \dot{X} = -\beta \int\limits_0^t d\tau \, \langle X(t-\tau)\dot{X}(0)\rangle x(\tau),$$

where use has been made of the relation $X(\Gamma(0))e^{-\hat{L}_0(t-\tau)} = X(\Gamma(t-\tau))$ (we recall that, for $x = 0$, $\partial_t \rho = -\hat{L}_0 \rho$, but $\dot{X} = \hat{L}_0 X$). The conclusion is that we have found a new (dynamical) fluctuation-dissipation relation

$$\langle X(t)\rangle_x = \int\limits_0^t d\tau \, \chi(t-\tau)x(\tau), \qquad \chi(t) = \beta \frac{d}{dt}\langle X(t)X(0)\rangle.$$

Notice, in analogy with the static case, that the fact that we have worked within a linear approximation, has the consequence that the correlation $\langle X(t)X(0)\rangle$ is calculated for $x = 0$. Further discussion of this subject can be found in [R. Zwangig, "Non-equilibrium Statistical Mechanics" (Oxford, 2001)].■

6.4 Temporal Reversibility

We have seen that the fluctuations of a macroscopic variable, z, in an isolated thermodynamic system, are associated with fluctuations of the Boltzmann entropy, $S(z)$: formation of a fluctuation is associated with a decrease of $S(z)$ (cooling of the system); dissipation of the fluctuation is associated with an increase of $S(z)$ (heating of the system). The heating-cooling of the system can be seen as an energy exchange between microscopic (thermal) degrees of freedom and macroscopic degrees of freedom (the fluctuations). From stationarity of the thermal equilibrium state, for an isolated system, the statistics of the fluctuations must itself be stationary.

More complicate the issue of time symmetry of the fluctuation statistics. With this we mean the condition that a certain history $z(\tau)$, in the interval $[0, t]$, of the thermodynamic fluctuation z, must have the same probability to be observed, as the process obtained by time reversal $\tilde{z}(\tau) = z(t - \tau)$. In other words, the following relation between realizations of the stochastic process $z(\tau)$ must be satisfied:

$$\rho[z] = \rho[\tilde{z}]. \tag{6.45}$$

It must be stressed that this condition is completely independent of that of stationarity (think of a saw-teeth signal, that is stationary, but certainly not time symmetric). The time symmetry of the fluctuations is in fact the product of two simultaneous effects:

- The stationary nature of thermal equilibrium.
- The time-symmetry of the microscopic dynamics of the isolated system, consequence of its Hamiltonian nature

The symmetry condition in Eq. (6.45) has an important realization, in terms of joint PDF's, known under the name of **detailed balance**:

$$\rho[z(t_1) = a, z(t_2) = b] = \rho[z(t_1) = b, z(t_2) = a]. \tag{6.46}$$

Notice that this condition does not imply at all that the PDF to pass from a to b, and that to pass from b to b, are equal. These are the conditional PDF's

$$\rho(z(t_2) = b | z(t_1) = a) \quad \text{and} \quad \rho(z(t_2) = a | z(t_1) = b),$$

and obviously, the PDF to pass from an equilibrium state to one of non-equilibrium, will be smaller than that corresponding to the inverse process. Detailed balance states a different condition: the PDF of observing a transition from a value of z at equilibrium to another in non-equilibrium must be equal to that of observing the reversed process (which is rather reasonable: for any departure from equilibrium, there must be a return to equilibrium; the detailed balance condition tells us simply that the return must take place in form identical to the departure).

6.4.1 Onsager Relations

The detailed balance condition has an important consequence as regards the way fluctuations are dissipated in thermal equilibrium. This condition can be translated, in terms of time correlations, into the following symmetry:

$$\langle z_i(0) z_j(t) \rangle = \langle z_j(0) z_i(t) \rangle,$$

that implies

$$\langle z_i(0) \dot{z}_j(t) \rangle = \langle z_j(0) \dot{z}_i(t) \rangle. \tag{6.47}$$

The generalization of the equation $\dot{z} = \gamma z + \xi$, to the case of many variables, is (sum over repeated indices understood):

$$\dot{z}_i = \gamma_{ij} z_j + \xi_i, \tag{6.48}$$

that, substituted into Eq. (6.47), gives us

$$\gamma_{jk} \langle z_i(0) z_j(t) \rangle = \gamma_{ik} \langle z_j(0) z_i(t) \rangle. \tag{6.49}$$

Taking the limit $t \to 0^+$ in Eq. (6.49), and exploiting the relation $\langle z_j z_i \rangle = (\alpha^{-1})_{ij}$, we find the symmetry

$$\gamma_{jk} (\alpha^{-1})_{ki} = \gamma_{ik} (\alpha^{-1})_{kj}.$$

Now, we can write the equation for the fluctuations (6.48) in the alternative form

$$\dot{z}_i = \lambda_{ij}\frac{\partial S}{\partial z_j} + \xi_i = -\lambda_{ij}\alpha_{jk}z_k + \xi_i, \tag{6.50}$$

where we have utilized the relation $S = \bar{S} - \frac{1}{2}\alpha_{ij}z_iz_j$ (see Eq. (6.3)). Comparing with Eq. (6.48), we find however $\lambda_{ij} = \gamma_{ik}(\alpha^{-1})_{kj}$, from which we obtain the following relation

$$\lambda_{ij} = \lambda_{ji}, \tag{6.51}$$

known as **Onsager relation**:

- The entropy decrease associated with the fluctuation z_j has the same effect of z_i, that the entropy decrease associated with the fluctuation z_i, would have on z_j.

The kinetic coefficients λ_{ij} have a second important property, that descends from the condition of entropy growth in the relaxation to equilibrium. From Eqs. (6.3) and (6.50), we can indeed obtain an evolution equation for the entropy. Taking an initial condition for S, with $\bar{S} - S$ much larger than the typical fluctuation at equilibrium, but sufficiently small for a linear approximation to be valid, the dynamics will be dominated by the relaxation

$$\dot{S} = \frac{\partial S}{\partial z_i}\dot{z}_i = \lambda_{ij}\frac{\partial S}{\partial z_i}\frac{\partial S}{\partial z_j}.$$

The condition of entropy growth in the relaxation to equilibrium, tells us that λ_{ij} is a matrix that is not only symmetric, but also positive defined.

Note It is not necessary to invoke a state of thermal equilibrium, in which the thermodynamic variables fluctuate near their equilibrium values, to have temporal reversibility. The microcanonical distribution takes into account the possibility, in principle, of fluctuations of arbitrary amplitude. The problem, as we have already discussed, is that such fluctuations would require a very long time to be observed. Ludwig Boltzmann, in his own days, hypothesized that the universe in which we live, and the time arrow that characterizes it, could be seen as the relaxation to equilibrium of a cosmic fluctuation. The initial condition for our universe would be the instant of maximum amplitude (and hence also minimum entropy) of the fluctuation. At this point, the reconciliation between a time-symmetric microscopic dynamics, and an irreversible macroscopic dynamic, would lie in the fact that, on the other side of the entropy minimum, there would be an arrow of time with direction opposite to ours. In other words, the evolution of the universe, in the view of Boltzmann, would consist of a sequence of unlikely fluctuations, near which, on both sides of the fluctuation maximum, a time arrow would be present, while for all the rest of the lifetime of the universe, a time arrow would not exist.■

6.4.2 Entropy Production

We have already seen that fluctuations change their nature in non-equilibrium conditions. They take a macroscopic character (turbulence) and become actually instru-
mental to the development of the relaxation process (see discussion in Sects. 3.3
and 3.10). This leads us to expect that, away from thermodynamic equilibrium,
fluctuations will loose their time-symmetric nature. In fact, the association between
breaking of time symmetry, and entropy production, is direct. In order to study the
connection, we focus on a stationary non-equilibrium situation, generated by means
of an external forcing on an otherwise isolated system. Entropy production will be,
in this case, simply the consequence of conversion of external work into heat.

We carry out the analysis with the simplest possible system: a Brownian particles
that is pushed around by a constant non-conservative force field. Let us indicate with
x the particle position, and assume for simplicity a linear drag force, proportional to
$\dot{\mathbf{x}}$. We have therefore the equation

$$M\ddot{\mathbf{x}} = -\gamma\dot{\mathbf{x}} + \mathbf{F}(\mathbf{x}) + \boldsymbol{\xi}, \qquad (6.52)$$

where $\boldsymbol{\xi}$ is the effect of molecular collision and M is the mass of the particle. We
consider an overdamped regime, such that the inertial term $M\ddot{\mathbf{x}}$ can be disregarded.
This corresponds to a situation in which the time scale for **x**, and therefore also for
F(x), is much longer than the relaxation time M/γ. In this regime, the inertia term
$M\ddot{\mathbf{x}}$ can be disregarded, and the dynamics reduces to

$$\gamma\dot{\mathbf{x}} = \mathbf{F}(\mathbf{x}) + \boldsymbol{\xi}. \qquad (6.53)$$

The heat transferred to the system, on the average, is just the mean work exerted by
the external force on the particle. The entropy production will be therefore

$$\Delta S = \frac{1}{T} \int_0^t d\tau \langle \mathbf{F}(\mathbf{x}(\tau)) \cdot \dot{\mathbf{x}}(\tau) \rangle d\tau. \qquad (6.54)$$

We see at once that, if the force is conservative, there will be no entropy production. In
fact, writing the force in terms of a potential, $\mathbf{F} = -\nabla V$, we would have, substituting
into Eq. (6.54):

$$S(t) - S(0) = -\frac{1}{T} \left\langle \int_0^t d\tau \, \langle \dot{\mathbf{x}}(\tau) \cdot \nabla V(\mathbf{x}(\tau)) \rangle \right\rangle = -\frac{1}{T} \langle [V(\mathbf{x}(t)) - V(\mathbf{x}(0))] \rangle,$$

that is equal to zero in stationary conditions. Comparing with Eq. (6.52), we notice
that, in presence of conservative forces, there is in fact a balance between entropy

production, $\gamma \dot{x}^2$ (fluctuation damping), and entropy destruction, $\xi \dot{x}$ (fluctuation production).

To study the relation with the time asymmetry of fluctuations in non-equilibrium, and entropy production, we focus on the properties of the PDF $\rho[\mathbf{x}|\mathbf{x}(0)]$, of the trajectories in the interval $[0, t]$, that have initial condition at $\mathbf{x}(0)$. In order to proceed, we discretize the equation of motion, Eq. (6.53), that becomes

$$\mathbf{x}_{k+1} - \mathbf{x}_k - \mathbf{F}_k \Delta \tau = \Delta \mathbf{B}_k, \tag{6.55}$$

where $\Delta \mathbf{B}(\tau_k) = \int_{\tau_k}^{\tau_{k+1}} \boldsymbol{\xi}(\tau) d\tau$, $\mathbf{x}_k \equiv \mathbf{x}(\tau_k)$, $\mathbf{F}_k = [\mathbf{F}(\mathbf{x}(\tau_{k+1})) + \mathbf{F}(\mathbf{x}(\tau_k))]/2$, and we have put for simplicity $\gamma = 1$. We notice at once, that, since $\boldsymbol{\xi}(\tau)$ is a white noise, the $\Delta \mathbf{B}(\tau_k)$'s at different times will be uncorrelated. As a consequence of this fact, $\mathbf{x}_k$ will be a Markov process. We notice also that $\Delta \mathbf{B}_k$ is a sum of infinite i.i.d. contributions, $\boldsymbol{\xi} d\tau$, so that we can expect the central limit theorem to be satisfied, and $\Delta \mathbf{B}_k$ to be Gaussian distributed. From isotropy of the molecular collisions, we expect the different components of $\Delta \mathbf{B}$ to be identically distributed, which gives us $\langle (\Delta B_i)^2 \rangle = D \Delta \tau$. We can write therefore

$$\rho(\Delta \mathbf{B}) \sim \exp\left(-\frac{|\Delta \mathbf{B}|^2}{2D\Delta t} \right). \tag{6.56}$$

Utilizing the discretized equation of motion (6.55), we can write the $\Delta \mathbf{B}(t_k)$'s that appear in $\rho[\Delta \mathbf{B}]$, as functions of $\mathbf{x}(t_k)$ and $\mathbf{x}(t_{k+1})$:

$$\rho(\mathbf{x}_{k+1}|\mathbf{x}_k) \sim \exp\left(-\frac{|\mathbf{x}_{k+1} - \mathbf{x}_k - \mathbf{F}_k \Delta \tau|^2}{2D\Delta \tau} - J_k \Delta \tau \right).$$

where J_n is the contribution from the Jacobian in the passage from $\rho_{\Delta \mathbf{B}}$ to $\rho_{\mathbf{x}_{n+1}}$.

Note The contribution from the Jacobian descends from the fact that we have utilized a "semi-implicit" definition for $\mathbf{F}_n$. We have here another example of rules of calculus that cease to work in the case of stochastic processes. Adopting as a definition, $\mathbf{F}_k = \mathbf{F}(\mathbf{x}(\tau_k))$, we would have obtained immediately $J_k = 0$, while, with $\mathbf{F}_k = [\mathbf{F}(\mathbf{x}(\tau_{k+1})) + \mathbf{F}(\mathbf{x}(\tau_k))]/2$, we find

$$\det\left[\frac{\partial \Delta \mathbf{B}_k}{\partial \mathbf{x}_{k+1}} \right] = \det\left[\delta_{ij} - \frac{\partial F_{k,i}}{\partial x_{k+1,j}} \Delta \tau \right] = 1 - (\nabla_{\mathbf{x}_{k+1}} \cdot \mathbf{F}_k) \Delta \tau + O((\Delta \tau)^2)$$

$$= \exp(-J_k \Delta \tau) + O((\Delta \tau)^2),$$

where $J_k = \nabla_{\mathbf{x}_{k+1}} \cdot \mathbf{F}_k = (1/2) \nabla \cdot \mathbf{F}(\mathbf{x}_{k+1})$. The "explicit" approximation $\mathbf{F}_k \simeq \mathbf{F}(\mathbf{x}(\tau_k))$ remains inaccurate also in the limit $\Delta \tau \to 0$.∎

To obtain the PDF of the history $\{x(t_k), k = 1, 2, \ldots\}$, we exploit the fact that $\mathbf{x}(t_k)$ is a Markov process:

$$\rho[\mathbf{x}] \simeq \rho(\mathbf{x}_n, \mathbf{x}_{n-1}, \dots, \mathbf{x}_0)$$
$$= \rho(\mathbf{x}_n|\mathbf{x}_{n-1})\rho(\mathbf{x}_{n-1}|\mathbf{x}_{n-2})\dots\rho(\mathbf{x}_1|\mathbf{x}_0)\rho(\mathbf{x}_0). \tag{6.57}$$

Substituting Eq. (6.56) into Eq. (6.57):

$$\rho[\mathbf{x}|\mathbf{x}(0)] \sim \exp\left[-\sum_k \left(\frac{|\mathbf{x}_{k+1} - \mathbf{x}_k - \mathbf{F}_k\Delta\tau|^2}{2D\Delta\tau} + J_k\Delta\tau\right)\right]. \tag{6.58}$$

In the same way, we can write the PDF of the history of the reversed process $\rho[\tilde{\mathbf{x}}|\tilde{\mathbf{x}}(0)]$, where $\tilde{\mathbf{x}}(\tau) = \mathbf{x}(t - \tau)$, so that the initial condition for the reverse process is the final condition for the direct process: $\tilde{\mathbf{x}}(0) = \mathbf{x}(t)$. Defining $\tilde{\mathbf{x}}_k = \mathbf{x}_{n-k}$ and $\tilde{\mathbf{F}}_k = [\mathbf{F}(\tilde{\mathbf{x}}_{k+1}) + \mathbf{F}(\tilde{\mathbf{x}}_k)]/2$:

$$\rho[\tilde{\mathbf{x}}|\tilde{\mathbf{x}}(0)] \sim \exp\left[-\sum_k \left(\frac{|\tilde{\mathbf{x}}_{k+1} - \tilde{\mathbf{x}}_k - \tilde{\mathbf{F}}_k\Delta\tau|^2}{2D\Delta\tau} + \tilde{J}_k\Delta\tau\right)\right]$$
$$= \exp\left[-\sum_k \left(\frac{|\mathbf{x}_{k+1} - \mathbf{x}_k + \mathbf{F}_k\Delta\tau|^2}{2D\Delta\tau} + J_k\Delta\tau\right)\right]. \tag{6.59}$$

Let us consider now the logarithm of the ratio of the PDF's of the two processes, direct and reversed, Eqs. (6.58) and (6.59). We find

$$\ln\left(\frac{\rho[\mathbf{x}|\mathbf{x}(0)]}{\rho[\tilde{\mathbf{x}}|\tilde{\mathbf{x}}(0)]}\right) = \frac{2}{D\Delta t}\sum_k \mathbf{F}_k \cdot (\mathbf{x}_{k+1} - \mathbf{x}_k)\Delta t$$
$$\simeq \frac{2}{D}\int_0^t \mathbf{F}(\mathbf{x}(\tau))\dot{\mathbf{x}}(\tau)d\tau = \frac{2Q}{D}, \tag{6.60}$$

where Q is the heat generated in the system from work of the force $\mathbf{F}$ along the trajectory $\mathbf{x}$. We assume that the trajectories are in statistical sense periodic:

$$\rho[\mathbf{x}(0) = \mathbf{x}_0] = \rho[\mathbf{x}(t) = \mathbf{x}_0].$$

This allows us to write $\rho[\mathbf{x}|\mathbf{x}(0)]\rho(\mathbf{x}(0)) = \rho[\mathbf{x}]$, and, at the same time, $\rho[\tilde{\mathbf{x}}|\tilde{\mathbf{x}}(0)]\rho(\mathbf{x}(0)) = \rho[\tilde{\mathbf{x}}|\tilde{\mathbf{x}}(0)]\rho(\tilde{\mathbf{x}}(0)) = \rho[\tilde{\mathbf{x}}]$. In these conditions, we find, from Eq. (6.60):

$$\ln\left(\frac{\rho[\mathbf{x}]}{\rho[\tilde{\mathbf{x}}]}\right) \simeq \frac{2Q}{D}, \tag{6.61}$$

and this expression will be different from zero in the case of non-conservative forces.

We have found an explicit example of the connection between non-conservative forces and violation of symmetry under time reversal: $\rho[\mathbf{x}] \neq \rho[\tilde{\mathbf{x}}]$. We find in particular the result that the trajectories associated with heat transfer to the system, are those whose probability is larger than that of their reversed in time companions. It is rather easy to understand what is happening, considering the example of a particle that is pushed along a circular trajectory by a non-conservative force. A typical trajectory **x** will correspond to the particle that travels around the circle in the same direction of the force, while $\tilde{\mathbf{x}}$ would be that extremely improbable trajectory, in which the molecular collisions push the particle in the direction opposed to the external force. The fact that the most probable trajectories are the ones associated with a positive heat release in the system, causes the average to RHS of Eq. (6.61), to be a positive quantity, equal to T times the entropy production in a cycle:

$$\langle \ln \rho[\mathbf{x}] \rangle - \langle \ln \rho[\tilde{\mathbf{x}}] \rangle = \frac{2T}{D}[S(t) - S(0)] \geq 0.$$

(We recall that all averages are calculated with respect to the forward PDF $\rho[\mathbf{x}]$).

6.5 Problems

Problem 1 A Brownian particle is connected by a spring to a solid support. In which way should the equation of motion for the particle be modified. Qualitatively describe the behavior of the particle, as a function of the spring constant, the particle mass, the drag by the fluid and the temperature. Determine the mean square distance of the particle from the support (work in one dimension).

Problem 2 Express the temperature fluctuation in a gas, in spatial Fourier components.

- Determine the fluctuation spectrum.
- Supposing the gas isolated, calculate the contribution of the fluctuations to the entropy (expand to second order in the fluctuations). We obtain an infinite result. Why? Comment the result.
- Write the evolution equation for the fluctuations in Fourier space, starting from the heat transport equation. Neglect the effect of compression and viscous heating in the gas.

Problem 3 A Brownian particle is placed in a Couette flow $\mathbf{u} = (\alpha y, 0, 0)$. Estimate the entropy production by the particle. Consider the fact that collisions push the particles in regions of the fluid in motion with respect to the original fluid parcel in which the particle was located. This means that more work will be required by the fluid to slow down the particle, with respect to the quiescent case.

Solution In a quiescent fluid, the mean power dissipated by the Brownian particle in the fluid is $(\gamma/2)\langle|\mathbf{v}|^2\rangle$, where γ is the drag coefficient, as indicated in Eq. (6.36). At thermal equilibrium, this same power is supplied back to the particle by molecular collisions through a term $\langle \boldsymbol{\xi} \cdot \mathbf{v} \rangle$, where $\boldsymbol{\xi}$ is the white noise term entering the same equation.

In the presence of the flow, the equation of motion for the particle must be modified to read

$$\dot{v}_x + \gamma(v_x - \alpha y) = \xi_x, \qquad \dot{v}_y + \gamma v_y = \xi_y, \qquad \dot{y} = v_y, \qquad (6.62)$$

while v_z remains uncoupled to the other components, and its evolution equation remains unmodified. It is convenient to redefine $\hat{\mathbf{v}} = \mathbf{v} - \mathbf{u}$, so that Eq. (6.62) becomes

$$\dot{\hat{v}}_x + \gamma\hat{v}_x + \alpha\hat{v}_y = \xi_x, \qquad \dot{\hat{v}}_y + \gamma\hat{v}_y = \xi_y, \qquad \dot{y} = \hat{v}_y. \qquad (6.63)$$

In terms of the new velocity, the power dissipated in the fluid, by the Brownian particle, will be

$$\langle\dot{Q}\rangle_\alpha = \frac{\gamma}{2}\langle|\hat{\mathbf{v}}|^2\rangle_\alpha.$$

We see that the power increment is all contained in the contribution from the x component:

$$\langle\dot{Q}\rangle_\alpha - \langle\dot{Q}\rangle_0 = \frac{\gamma}{2}[\langle\hat{v}_x^2\rangle_\alpha - \langle\hat{v}_x^2\rangle_0].$$

We calculate $\langle\hat{v}_x^2\rangle_\alpha$ from the first of Eq. (6.63). We have

$$\hat{v}_x(t) = \int_{-\infty}^{t} d\tau\, e^{-\gamma(t-\tau)}[\alpha\hat{v}_y(\tau) + \xi_x(\tau)].$$

From here we obtain

$$\langle\hat{v}_x^2\rangle_\alpha = \frac{\gamma}{2}\int_{-\infty}^{0} d\tau d\tau'\, e^{\gamma(\tau+\tau')}[\alpha^2\langle\hat{v}_y(\tau)\hat{v}_y(\tau')\rangle + D\delta(\tau - \tau')], \qquad (6.64)$$

where we have exploited the fact that $\langle\xi_x\xi_y\rangle = 0$ and therefore also $\hat{v}_y\xi_x = 0$. We recognize in the second term to RHS of Eq. (6.64), the contribution that would be present in the quiescent fluid. We find therefore

$$\langle\dot{Q}\rangle_\alpha - \langle\dot{Q}\rangle_0 = \frac{\gamma}{2}\int_{-\infty}^{0} d\tau d\tau'\, e^{\gamma(\tau+\tau')}\alpha^2\langle\hat{v}_y(\tau)\hat{v}_y(\tau')\rangle.$$

Exploiting $\langle\hat{v}_y(t)\hat{v}_y(0)\rangle = \langle\hat{v}_y^2\rangle e^{-\gamma|t|}$, we find

$$\langle\dot{Q}\rangle_\alpha - \langle\dot{Q}\rangle_0 = \frac{\alpha^2}{4\gamma}\langle\hat{v}_y^2\rangle, \qquad (6.65)$$

where, from Eq. (6.63), $\langle \hat{v}_y^2 \rangle$ has the same value as in the quiescent fluid. Notice that presence of the flow, does not modify the forcing term ξ, that is associated solely with molecular effects. The whole increase in the fluctuation amplitude for $\hat{v}$, with respect to the quiescent case, is produced by the flow. The difference $\langle \dot{Q} \rangle_\alpha - \langle \dot{Q} \rangle_0$ is therefore the heating in the system produced by the shuffling around of the Brownian particle by $\mathbf{u}$. Dividing Eq. (6.65) by T, we obtain the entropy production in the fluid.

6.6 Further Reading

Basic references on the issues covered in this chapter:

- R. Zwangig, *Non-equilibrium Statistical Mechanics* (Oxford, 2001)
- K. Sekimoto, *Stochastic Energetic* (Springer, 2010)

The section on the connection between irreversibility and entropy production is based on:

- P. Gaspard, Time-reversed dynamical entropy and irreversibility in Markovian random processes, J. Stat. Phys. **117**, 599 (2004)

Additional material on connections between stochastic processes, field theory, and break-up of the rules of standard calculus can be found in:

- R.P. Feynman, A.R. Hibbs, *Quantum Mechanics and Path Integrals* (McGraw Hill, 1965)
- L.S. Schulman, *Techniques and applications of path integration* (Wiley and Sons, 1996)

and again in

- P.E. Kloeden, E. Platen, *Numerical Solution of Stochastic Differential Equations* (Springer, 1992)

Index

© Springer International Publishing Switzerland 2015
P. Olla, *An Introduction to Thermodynamics and Statistical Physics*,
UNITEXT for Physics, DOI 10.1007/978-3-319-06188-7

Printed in the United States
By Bookmasters